Sustainable Civil Infrastructures

About this Series

Sustainable Infrastructure impacts our well-being and day-to-day lives. The infrastructures we are building today will shape our lives tomorrow. The complex and diverse nature of the impacts due to weather extremes on transportation and civil infrastructures can be seen in our roadways, bridges, and buildings. Extreme summer temperatures, droughts, flash floods, and rising numbers of freeze-thaw cycles pose challenges for civil infrastructure and can endanger public safety. We constantly hear how civil infrastructures need constant attention, preservation, and upgrading. Such improvements and developments would obviously benefit from our desired book series that provide sustainable engineering materials and designs. The economic impact is huge and much research has been conducted worldwide. The future holds many opportunities, not only for researchers in a given country, but also for the worldwide field engineers who apply and implement these technologies. We believe that no approach can succeed if it does not unite the efforts of various engineering disciplines from all over the world under one umbrella to offer a beacon of modern solutions to the global infrastructure. Experts from the various engineering disciplines around the globe will participate in this series, including: Geotechnical, Geological, Geoscience, Petroleum, Structural, Transportation, Bridge, Infrastructure, Energy, Architectural, Chemical and Materials, and other related Engineering disciplines.

More information about this series at http://www.springer.com/series/15140

Manik Barman · Musharraf Zaman
Jia-Ruey Chang
Editors

Transportation and Geotechniques: Materials, Sustainability and Climate

Proceedings of the 5th GeoChina International Conference 2018 – Civil Infrastructures Confronting Severe Weathers and Climate Changes: From Failure to Sustainability, held on July 23 to 25, 2018 in HangZhou, China

Springer

Editors
Manik Barman
Department of Civil Engineering
University of Minnesota Duluth
Duluth, MN, USA

Jia-Ruey Chang
Graduate Institute of Architecture
National Ilan University
Yilan, Taiwan

Musharraf Zaman
School of Civil Engineering
University of Oklahoma
Norman, OK, USA

ISSN 2366-3405 ISSN 2366-3413 (electronic)
Sustainable Civil Infrastructures
ISBN 978-3-319-95767-8 ISBN 978-3-319-95768-5 (eBook)
https://doi.org/10.1007/978-3-319-95768-5

Library of Congress Control Number: 2018948642

Printed on acid-free paper

This Springer imprint is published by the registered company Springer International Publishing AG part of Springer Nature
The registered company address is: Gewerbestrasse 11, 6330 Cham, Switzerland

Contents

Introduction

The technical papers presented in this volume deal with the attempts made by the scientists, researchers, and practitioners to address different emerging issues in transportation and geotechnical engineering. A total of 15 technical papers are included in this volume covering a wide spectrum of transportation and geotechnical issues, such as (i) unpaved road, (ii) airport pavements, (iii) pavement preservation techniques, (iv) metro stations, (v) bridge approach, (vi) warm-mix asphalt, (vii) cement concrete materials, (viii) behaviour of pile foundation, and (ix) seismic liquefaction. This volume is part of the proceedings of the 5th GeoChina International Conference on Civil Infrastructures Confronting Severe Weathers and Climate Changes: From Failure to Sustainability, HangZhou, China 2018.

Evaluation of Polymer Based Dust Suppressant Mixed with Clayey Soil in Unpaved Road: Lab Experiment

Junxin Huang[1], Chun-Hsing Ho[2(✉)], Yijie Gao[2], Ziyan Wu[2], and Yuzhi Zhang[2]

[1] School of Resource Environment and Safety Engineering, Hunan University of Science and Technology, Xiangtan 411201, China
15886425576@126.com
[2] Department of Civil Engineering, Construction Management and Environmental Engineering, Northern Arizona University, PO Box 15600, Flagstaff, AZ 86011, USA
{chun-hsing.ho,yijiegao,zw222,yz97}@nau.edu

Abstract. Industrial activities, combustion, roads and construction are all sources for the generation of fugitive dust. Fugitive dust presents a major challenge for air quality, unpaved road preservation and transportation safety. In the past, a variety of dust suppressants was used in the industry to cope with this problem. In this project, a polymer-based suppressant was analyzed to see its performance on clayey soil. The clayey soil samples were collected in Sedona, Arizona and were mixed with polymer-based suppressant with 3 concentration rates, 5%, 10%, and 15% by weight. A series of lab experiments were performed including moisture retention tests, surface strength tests, dynamic rolling (grinding) tests, and scanning electron microscopy (SEM) imaging. The mechanical responses data of the dust suppressant treated specimens were recorded and analyzed. The results show that the 15% polymer treated specimen has a better ability in dust suppression as shown in its moisture retention rate, surface strength/resistance to load penetration, dynamic rolling resistance, and cohesion/interlock effect in SEM images.

1 Introduction

Today there are more than 400 million miles of roads in the United States; approximately 50% of them are gravel roads (unpaved) (Wang and Wang 2014). These roads are often subject to dusting due to traffic loading and the effects of natural environmental factors (i.e., wind, storm, etc.). Fugitive dust will reduce air quality and road visibility. There are a number of direct methods to be used for dust control and mitigation, such as construction and maintenance, mechanical stabilization and chemical palliatives (Bai et al. 2005). The "dust" can be defined simply as particles of soil that have become airborne. In general, dust is mainly due to particles in the soil less than 0.075 mm in diameter, that is, through the #200 sieve particles can produce dust effects (Cooper and Alley 2011). To date, the most effective and most convenient way is to use chemical reagents to control dust. Nowadays in the United States, many local

M. Barman et al. (eds.), *Transportation and Geotechniques: Materials, Sustainability and Climate*, Sustainable Civil Infrastructures,
https://doi.org/10.1007/978-3-319-95768-5_1

governments apply various dust suppressants on their roadway, mainly calcium chloride and magnesium chloride. This paper used polymer suppressant product as a dust control method obtained from a local company in Phoenix, Arizona, USA. The soil sample was collected from Sedona located in northern Arizona of the U.S.

Four tests were used to evaluate the mechanical responses of polymer-based suppressant in mitigating fugitive dusts: Moisture Retention Test, Surface Strength Tests, Dynamic Rolling Tests and Scanning Electron Microscopy Analysis.

2 Sample Preparation

As previously mentioned, polymer suppressant product was obtained from Desert Mountain Inc. to be used in the research. The soil sample was collected from Sedona. Polymer suppressant solutions were dissolved and reproduced with 5%, 10%, and 15% concentrations by weight for the experiment. Figure 1 shows the particle size distribution of the soil sample. The figure also shows that 90.92% of the material passed the #4 (4.75 mm) sieve and 1.19% was finer than sieve #200 (0.075 mm). The uniformity coefficient and the coefficient of curvature were calculated to be 5.91 and 0.44. The sample soil type is A-3 (fine sand) based on the AASHTO method. It was determined to be a poorly-graded sand with clay (or silty clay, SP-SC) based on the Unified Soil Classification System (USCS) method compatible with ASTM D2487, D2488 standards and loamy sand based on the United States Department of Agriculture (USDA) method.

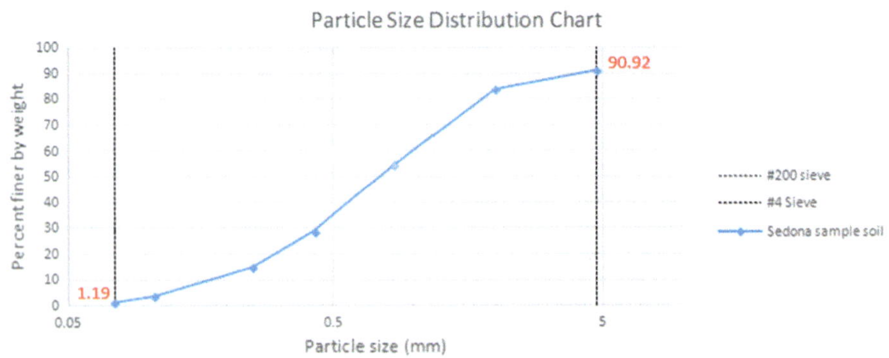

Fig. 1. Particle size distribution of soil sample

The typical application rate of dust suppressant in practice varies from 0.5 to 4.5 L/m2 depending on the dust suppressant and the conditions of the site (Bolander and Yamada 1999). In the mixing process, 50 mL of water, 5%, 10%, and 15% of polymer solutions were slowly sprayed over the surface of the four soil samples with an application rate of 1.59 L/m^2 until the four specimens were fully mixed with the polymer suppressant. It is noticed that of the four different polymer suppressant solutions, specimens mixed with water (0%, control group), 5%, and 10% polymer suppressant solution were able to be quickly penetrated into soil particles. However, 15% of polymer solution could not fully mix with the soil sample without manual stirring. This type of polymer treatment solution (15%) exhibits adverse workability in mixing with soil sample. For construction purposes, the optimum application rate for polymer suppressant treatment should be determined less than 15%.

3 Test Results and Discussions

3.1 Moisture Retention Tests

In this project, the moisture retention capacity of four polymer treated specimen is studied. Moisture change of the four specimens before and after air dry condition is shown in Fig. 2. As can be seen, the moisture content of four soil specimens dropped as time increased. After 4 h, all moisture changes tend to be less significant. The soil specimen mixed with 15% of polymer suppressant solution exhibits highest moisture retention capacity than the other specimens. This finding indicates the polymer suppressant has better capacity in keeping moisture within soil paste because of its strong bonding capacity. The enhancement of moisture retention capacity is beneficial to dust control. Not only it helps with bonding soil particles firmly, but it also increases the service duration of dust suppressant so as to reduce the operation cost. On the other hand, this test could also show this type of polymer has the ability to form soil particles together if some roadway does not need dust control. In addition, this polymer could help the roadway keep steady instead of becoming muddy after the heavy snowstorm or rainstorm in some areas in Arizona.

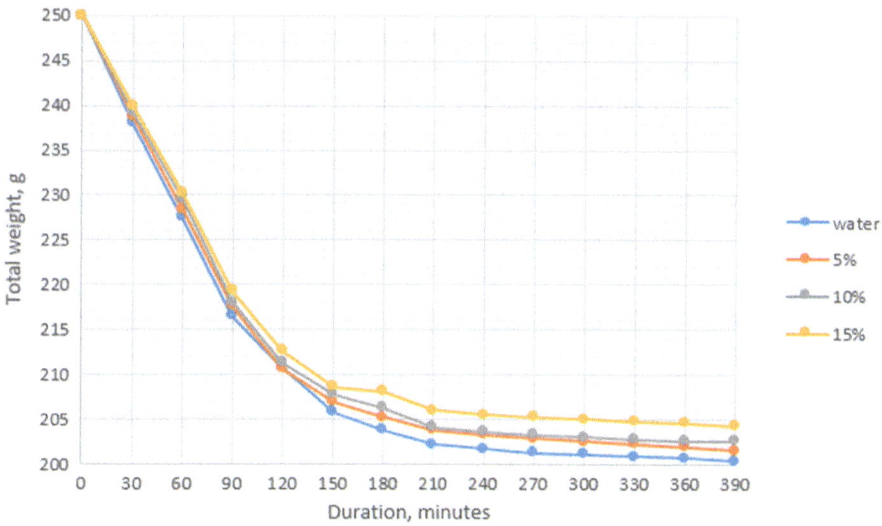

Fig. 2. Moisture changes versus time duration

3.2 Surface Strength (Penetration Resistance) Test

Fugitive dust movement depends on how easily the soil particles can be detached from the surface. This is measured by the surface strength of the soil. Researchers have used different methods to estimate the surface strength of the soil, including the modulus of rupture method (Richards 1953; Gillette et al. 1982), the needle penetrometer method (Bengough and Mullins 1990; Rice et al. 1997), the fall cone or cone penetrometer (Bradford and Grossman 1982; Campbell and O'Sullivan 1990), the torvane method (Govers and Poesen 1986), and the aggregate stability method (Skidmore and Powers 1982). Chen et al. (2015) provided a brief review and evaluation of these different methods and recommended a simple flat-ended cylindrical penetrometer method for estimating the surface strength of the soil. This method provides better analysis results than other methods because it applies a load to a surface area encompassing many grains rather than just a few grains as using a needle penetrometer.

In this study, the method recommended by Chen et al. (2015) was adopted to assess the surface strength of all soil specimens treated with different polymer solutions. Before penetrometer test, all soil specimens were placed in an oven for 24 h and then removed and placed on the table. Ten penetrometer tests were randomly applied on the surface of a soil specimen. Based on the soil resistance, 10 penetrated/fractured areas on the surface were examined and readings were recorded (Fig. 3). The penetrometer test results are shown in Fig. 4.

As shown in Fig. 4, the soil specimen treated with the higher polymer soil concentration shows higher surface resistance and their surrounding penetrated areas did not exhibit substantially crashed by meaning of soil particles were being bonded firmly. Figure 5 provides 10 reading values and their average for the four soil specimens. Obviously, the specimen treated with 15% polymer position has the highest surface strength (penetration resistance) with a mean of 2034 kPa. All statistical data were depicted in Table 1. The surface strength did not increase substantially (difference = 419 kPa) between the 10% polymer treated specimen (mean = 1615 kPa) and 15% polymer treated specimen. The less significant difference between the two treated specimens (10% and 15%) would give users an option to adopt a dust suppressant with 10% of polymer solution as the 15% polymer treated dust suppressant does not show better workability and that is too sticky to spray over the road surface.

3.3 Dynamic Rolling (Grinding) Test and Discussion

In order to simulate the effect of dynamic traffic loading on the performance of polymer treated specimens, a dynamic rolling (grinding) test was used. A total of 12 rubber bunch balls (3.3 cm in diameter, each) were placed on the surface of a soil specimen covered by a 3-kg weight mass (18 cm in diameter). The entire set was then located on a mechanical shaker running/shaking for 6 min. During shaking, rubber bunch balls were able to roll over the surface of a soil specimen with a means to simulate dynamic vehicle travelling on road surface. After completion of a dynamic rolling test, the remaining soil was used to perform a sieve analysis and its particle distribution curve was created and compared with the original one. Figure 5 exhibits five particle distribution curves including the four polymer treated specimens and the original soil

(a) specimen treated with water

(b) specimen treated with 5% pine oil

(c) specimen treated with 10% pine oil

(d) specimen treated with 15% pine oil

Fig. 3. Surface conditions of penetrometer tests

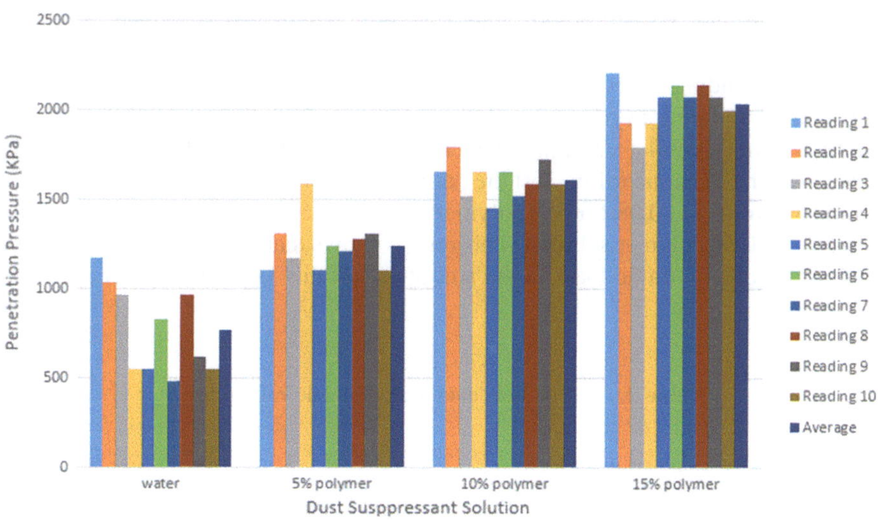

Fig. 4. Surface strength of soil specimens

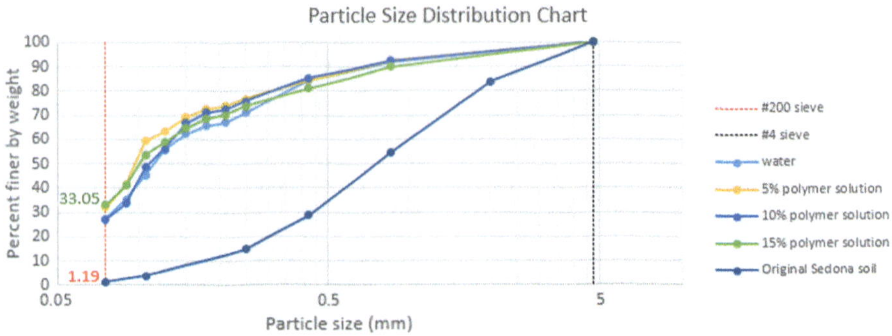

Fig. 5. Soil distribution curves before and after dynamic rolling test

Table 1. Statistics of surface strength of four specimens

Polymer Solution	Mean, KPa	Standard Deviation, KPa	COV
Water	773	236.5	30.6%
5% Polymer Solution	1242	138.6	11.2%
10% Polymer Solution	1615	98.6	6.1%
15% Polymer Solution	2036	116.5	5.7%

sample. The particles passing #200 sieve from the four specimens are calculated and are shown in Table 2. The particles passing #200 sieve is an indication that shows how easily fugitive dust can be generated based on vehicle travelling/traffic loading on road surface. Obviously, the soil treated with moisture shows the least resistance (27.36% of particles passing #200) to dynamic rolling (traffic loading) followed by the 5% polymer treated specimen (32.01% of particles passing #200), 10% polymer treated specimen (27.06% of particles passing #200), and 15% polymer treated specimen (33.06% of particles passing #200). Based on this test, the finding helps better understanding the interlock and bonding effect of the soil particles with respect to the levels of polymer application. Comparing to original soil sieve analysis data, the particles size between #200 (0.075 mm) to #4 (4.75 mm) percent is 98.81 (difference between 100% and 1.19%), after the dynamic rolling test the particles size between #200 (0.075 mm) to #4 (4.75 mm) percent is 60.08 for 5% polymer solution, 65.45 for 10% polymer solution and 57.04 for 15% polymer solution. This results shows that the clay soil particles have deeply influenced by traffic load on the soil after add this kind of polymer solution in

Table 2. Soil particle data after dynamic rolling test

Soil specimens with treatment	Water, 0%	5% polymer	10% polymer	15% polymer	Original sample
% Passing #200	27.36	32.06	27.06	33.05	1.19

the macro view because of amount of the soil particles retain large diameter particles was less than original soil sample.

3.4 Scanning Electron Microscopy Experiment

A scanning electron microscopy (SEM) instrument was employed in the research work. After completion of surface strength tests using a pocket penetrometer, SEM samples were collected from each specimen that before the surface strength test and dynamic rolling test. The scanning electron microscope scans a focused electron beam over the surface of a pine soil treated specimen to create an image in which the microstructure of the specimen was able to be examined and measured. The SEM imaging was performed using a Zeiss Supra 40VP SEM located at Northern Arizona University. During SEM imaging, specimens were zoomed to 215x to allow measuring dimensions and observing soil particle interlock and bonding situations with respect to the levels of polymer treatment. Figure 6 exhibits SEM images with 215x. In the specimen with moisture treatment, the gap between soil particles has not obviously change of 5% polymer treated specimens. However, as can be seen in Fig. 6, polymer suppressant created thin films that bond and cover particles to increase their cohesion as well as interlock effect, particularly in the 15% polymer treated specimen in Fig. 6(d).

(a) Specimen with water

(b) Specimen with 5% polymer treatment

(c) Specimen with 10% polymer treatment

(d) Specimen with 15% polymer treatment

Fig. 6. SEM 215x images of four specimens

Based on the above experiments, 15% of polymer solution seemed to have had better abilities to be used in dust suppression, provided its moisture retention rate, surface strength/resistance to load penetration, and cohesion/interlock effect in SEM images, these three parts results show promising results than other specimens.

4 Conclusions and Recommendations

This paper presents a lab experiment approach to quantify mechanical responses of polymer dust suppressant mixed with native soil in Northern Arizona. Through a series of lab experiments, the following conclusions are drawn:

1. Polymer-based specimens exhibit better moisture retention abilities than specimens treated with water. The moisture retention rate increases as the concentration of polymer solution increases.
2. Polymer-based specimens show better surface strength than specimens treated with water. The surface strength and resistance to load penetration increase as the concentration of polymer solution increases.
3. After dynamic rolling tests, the soil treated with moisture shows the least resistance to dynamic rolling (traffic loading) followed by the soil sample treated with 5%, 10%, and 15% polymer-based solutions. Soil mixed polymer-based dust suppressant does show better resistance to vehicle travelling.
4. As observed in SEM images, polymer-based solutions suppressant created thin films that could bond and cover particles to increase cohesion and interlock effect that contributed to the decrease in generation of fugitive dust during construction, mining, and vehicle travelling on unpaved roads.
5. The concentration of polymer suppressant controls the workability in the mixing with soil particles. Based on lab experiments, 5% and 10% of polymer-based solutions could fully spray over the surface of a soil sample, and penetrate through soil particles. However, 15% of polymer-based solutions could not fully mix with soil sample without manual stirring. Most of 15% polymer-based solutions solution liquids stayed on the surface of a soil sample. This type of polymer-based solutions solution (15%) does exhibit adverse workability in mixing with soil sample. For construction purposes, the optimum solution rate should be selected in between 10% to 15% to obtain a better pine oil suppressant application in the field.

Acknowledgements. This research work is supported by the following grants: (1) Natural Science Foundation of Hunan Province, China (18JJ4028), (2) the Foundation of National Key Laboratory for Safety and Health of Mines sponsored by the MaanShan Institute of Mining Research CO., LTD., SinaSteel Group, China (2017-JSKSSYS-05), (3) the Industrial Collaboration Funding of Hengyang Science and Technology Program, China (2015kc58), (4) 2016 Technology Funding for Safe Production and Severe Incident Prevention (Hunan-0003-2016AQ), and (5) 2017 Technology Funding for Safe Production and Severe Incident Prevention (Hunan-0007-2017AQ), and (6) the Opening Foundation Project of Hunan Provincial Key Laboratory of Safe Mining Techniques of Coal Mines at Hunan University of Science and Technology, China (201405). The authors expressed their gratitude for the above funding supports. The authors also would like to thank Desert Mountain Inc. for providing polymer dust suppressants to be used in the experiments. The assistance from staff of Desert Mountain Inc. is gratefully appreciated.

References

Bai, X.B., Liu, J., Yan, I.T.: Fugitive dust pollution of city and situation and prospect of study on dust-depressor. J. Shaanxi Univ. Technol. **21**(4), 43–46 (2005). (In Chinese)

Bengough, A.G., Mullins, C.E.: Mechanical impedance to root growth: a review of experimental techniques and root growth responses. J. Soil Sci. **41**(3), 341–358 (1990)

Bolander, P., Yamada, A.: Dust Palliative Selection and Application Guide. Project Rep. 9977-1207-SCTDC, U.S. Dept. of Agriculture, Forest Service, San Dimas Technology and Development Center, San Dimas, CA (1999)

Bradford, J.M., Grossman, R.B.: In-situ measurement of nearsurface soil strength by the fall-cone device. Soil Sci. Soc. Am. J. **46**(4), 685–688 (1982)

Campbell, D.J., O'Sullivan, M.F.: The cone penetrometer in relation to trafficability, compaction, and tillage. In: Smith, K., Mullins, C.E. (eds.) Soil Analysis: Physical Methods, pp. 399–429. Marcel Dekker, New York (1990)

Chen, R., Lee, I., Zhang, L.Y.: Biopolymer stabilization of mine tailings for dust control. J. Geotech. Geoenviron. Eng. **141**(2), 04014100-1–04014100-100 (2015)

Cooper, C.D., Alley, F.C.: Air Pollution Control-A Design Approach, 4th edn. Waveland Press, Incorporated (2011)

Gillette, D.A., Adams, J., Muhs, D., Kihl, R.: Threshold friction velocities and rupture moduli for crusted desert soils for the input of soil particles into the air. J. Geophys. Res. **87**(C11), 9003–9015 (1982)

Govers, G., Poesen, J.: Afield-scale study of surface sealing and compaction of loam and sandy loam soils. Part 1. Spatial variability of surface sealing and crusting. In: Callebaut, F., Gabriels, D., DeBoodt M (eds.) Proceedings of the Symposium on Assessment of Soil Surface Sealing and Crusting, pp. 171–182. Flanders Research Centre for Soil Erosion and Soil Conservation, Ghent, Belgium (1986)

Rice, M.A., Mullins, C.E., McEwan, I.K.: An analysis of soil crust strength in relation to potential abrasion by saltating particles. Earth Surf. Process. Land. **22**(9), 869–883 (1997)

Richards, L.A.: Modulus of rupture as an index of crusting of soil. Soil Sci. Soc. Am. J. **17**(4), 321–323 (1953)

Skidmore, E.L., Powers, D.H.: Dry soil-aggregate stability: energy-based index. Soil Sci. Soc. Am. J. **46**(6), 1274–1279 (1982)

Wang, X., Wang, W.J.: Urban dust pollution and its research of control method in China. Environ. Sci. Technol. **37**(120), 588–592 (2014). (In Chinese)

Evaluation of Co-polymer Based Dust Suppressant Mixed with Soil in Unpaved Road: Lab Experiment

Junxin Huang[1], Chun-Hsing Ho[2(✉)], Ziyan Wu[2], Yijie Gao[2], Yuzhi Zhang[2], and Zhonghan Zhang[2]

[1] Hunan Key Laboratory of Safe Mining Techniques of Coal Mines, Hunan University of Science and Technology, Xiangtan 411201, China
15886425576@126.com

[2] Department of Civil Engineering, Construction Management and Environmental Engineering, Northern Arizona University, PO Box 15600, Flagstaff, AZ 86011, USA
{chun-hsing.ho, ziyanwu, yijiegao, yuzhizhang, zz67}@nau.edu

Abstract. Fugitive dusts, which are considered as a type of the atmospheric particular matter (PM), are produced from the surface of unpaved roads. Since fugitive dusts are atmospheric PM, they can cause respiratory illness and lung damage for humans, and even lead to premature death in sensitive individuals. In order to reduce the fugitive dusts generated from the road surface, a number of dust suppressants have been tested and used in the construction industry. In this paper, a co-polymer based suppressant was used, aiming to find a good method to cope with the fugitive dust emissions. The three concentration rates of co-polymer based dust suppressants were prepared: 1%, 3%, and 5% by weight. The soil was sampled in Sedona, Arizona, USA, and was mixed with co-polymer based dust suppressants. A series of lab experiments were conducted including moisture retention tests, surface strength tests, dynamic rolling (grinding) tests, and scanning electron microscopy (SEM) imaging. Based on test results, the 5% co-polymer treated specimens showed better performance in mitigating the generation of fugitive dust emissions as compared with 0%, 1% and 3% co-polymer treated specimens.

Keywords: Dust suppressants · Dust control · Fugitive dusts · Co-polymer based solution

1 Introduction

In the United States, approximately 50% of all types of road (over 400 million miles) are unpaved (Wang and Wang 2014). According to air quality reports from Environmental Protection Agency (EPA), unpaved roads generated nearly 65% of the airborne dusts nationwide (Sanders et al. 2015; EPA 2006). Airborne dusts including fugitive dust emissions have led to the health tread of respiratory system, economy loss of the environmental departments, and equipment damage of machines on the road (Bai et al.

M. Barman et al. (eds.), *Transportation and Geotechniques: Materials, Sustainability and Climate*, Sustainable Civil Infrastructures,
https://doi.org/10.1007/978-3-319-95768-5_2

2005; Xi et al. 2017). Since approximately 50% of the roads in the United States produce nearly 65% airborne dusts, creating many negative impacts on the environment and humans, the dust control of the unpaved roads therefore is essential to protect the air quality and human health.

The diameter of fugitive dust is defined by the International Organization (ISO) for Standardization as the dusts whose diameter are less than 75 μm. Specifically, fugitive dusts are particular matters (PM) whose diameters are not greater 10 μm (PM 10) and PM 2.5 is referred as fugitive dusts are particular matters (PM) whose diameters are not greater 2.5 μm (Midwest Research Institute 2006). The above information indicates that the particles going through the #200 sieve can be regarded as fugitive dusts that would have a significant impact on the environment and humans (Cooper and Alley 2011). In the past, numerous types of dust suppressant are tested and used to control the fugitive dust emissions in the construction industry, such as water, aqueous industrial waste, and chemical dust suppressant. A co-polymer based dust suppressant is a part of chemical treatments used in road construction, particularly in unpaved roads. The paper is presented to evaluate the effectiveness of the co-polymer dust suppressants in mitigation of fugitive dusts. The soil was sampled in Sedona, Arizona, USA to be mixed with the co-polymer dust suppressants.

Four tests were performed in the Materials Laboratory of Northern Arizona University to estimate the mechanical responses of co-polymer-based suppressants in reducing the generation of fugitive dust emissions: Moisture Retention Test, Surface Strength Tests, Dynamic Rolling Tests, and Scanning Electron Microscopy Analysis.

2 Material Preparation

Surface-Loc suppressant, a co-polymer dust suppressant product, is obtained from a supplier in Phoenix, AZ, USA, and it was mixed with the soil collected from Sedona, Arizona. Figure 1 shows the particle size distribution of the soil sample. The figure also shows that 90.92% of the material passed the #4 (4.75 mm) sieve and 1.19% was finer than sieve #200 (0.075 mm). The uniformity coefficient and the coefficient of curvature of the soil were calculated to be 5.91 and 0.44. The soil type is categorized as A-3 (fine sand) based on the AASHTO method. In addition, the soil was determined to be a poorly-graded sand with clay (or silty clay, SP-SC) using the Unified Soil Classification System (USCS) method compatible with ASTM D2487, D2488 standards, and loamy sand based on the United States Department of Agriculture (USDA) method.

The co-polymer suppressant solutions were dissolved and reproduced with 0% (mixed with water as a control group) 1%, 3%, and 5% concentrations by weight for the experiment. The typical application rate of dust suppressant in practice varies from 0.5 to 4.5 L/m^2 depending on the dust suppressants and the conditions of the site (Bolander and Yamada 1999). Soils passing #200 were collected for the experiment. During the producing process, four different co-polymer suppressant solutions were prepared, and they were slowly sprayed over the surface of the soil samples with an application rate of 1.59 L/m^2 until the specimens were fully mixed with the co-polymer suppressants.

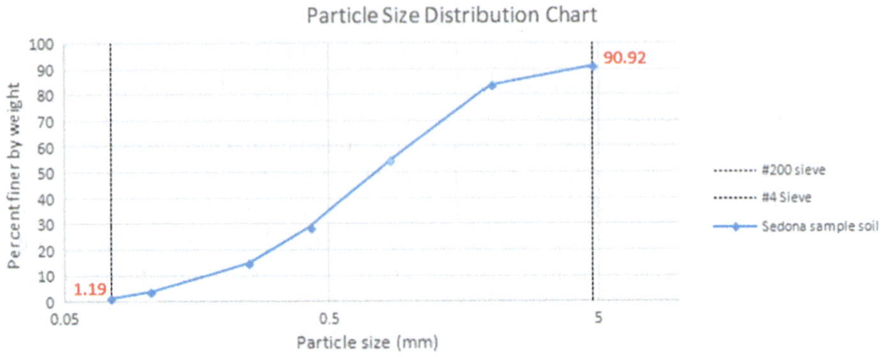

Fig. 1. Particle size distribution of soil sample

3 Test Results and Discussions

3.1 Moisture Retention Tests

The ability of keeping moisture within co-polymer suppressant treated soils is a critical factor that shows how well the suppressants can keep the dust emissions from being taken away from the road surface. The moisture retention tests aimed to evaluate the moisture capacity of the co-polymer suppressants under different concentrations (0%, 1%, 3%, and 5%). Four different soil specimens with the same total weight (250 g, 200 g of soil sample and 0 g of a designed co-polymer solution) but different concentrations were prepared. The four specimens were placed outsides the laboratory and exposed to the air and sunshine. The condition in which the specimens was exposed was similar with the real practice in the unpaved roads. The specimens were weighed in a 30-min interval in the laboratory to appraise the moisture content deduction over a period of 180 min. The results are shown in Fig. 2.

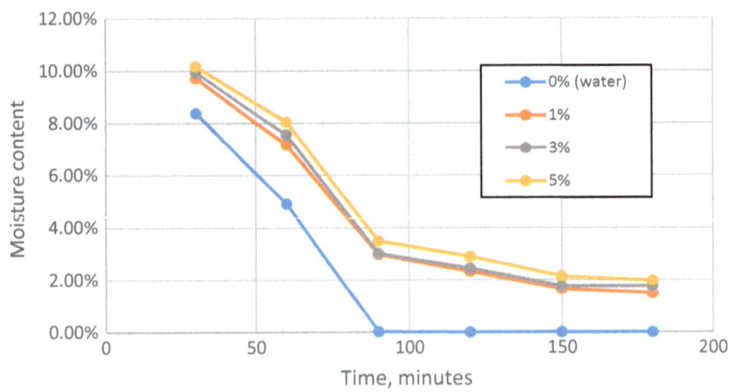

Fig. 2. Moisture changes versus time duration

As shown in Fig. 2, the moisture contents of the four specimens all decrease as the time increases. However, after 1.5 h (90 min) the decreasing rate of the moisture contents tend to be less significant. In addition, the specimen with 5%-concentration has the highest moisture retention capacity as compared with the others. The enhancement of the moisture retention capacity helps with bonding soil particles firmly, and also increases the service life cycles of co-polymer treated unpaved roads which reduces the operation cost.

3.2 Surface Strength (Penetration Resistance) Test

Fugitive dust emission movement depends on how easily the soil particles can be detached from the surface. Many methods have been used to measure the surface strength of the soil specimens treated by dust suppressants, comprising the modulus of rupture method (Richards 1953; Gillette et al. 1982), the needle penetrometer method (Bengough and Mullins 1990; Rice et al. 1997), the fall cone or cone penetrometer (Bradford and Grossman 1982; Campbell and O'Sullivan 1990), the torvane method (Govers and Poesen 1986), and the aggregate stability method (Skidmore and Powers 1982). Chen et al. (2015) provided a brief review and evaluation of these different methods and recommended a simple flat-ended cylindrical penetrometer method for estimating the surface strength of the soil. This method provides better analysis results than other methods because it applies a load to a surface area encompassing many grains rather than just a few grains using a needle penetrometer. In this study, the method recommended by Chen et al. (2015) was used to assess the surface strength of all soil specimens treated with different co-polymer solutions. Before a penetrometer test, all soil specimens were placed in an oven for 24 h and then removed and placed on a table. Ten penetrating tests were randomly applied on the surface of a soil specimen. Based on the soil resistance, 10 penetrated/fractured areas on the surface were examined and readings were recorded (Fig. 3), and the results are shown in Fig. 4.

As shown in Fig. 4, the soil specimens treated with the higher co-polymer soil concentration show higher surface resistance and their surrounding penetrated areas did not exhibit substantially crashed, provided that soil particles were being bonded firmly. During testing, it was noted that the specimens treated with 5% co-polymer solutions were too strong to penetrate by means of exceeding the maximum capacity of the device (17,926 kPa). Thus, the 17,926 kPa of ultimate reading was used for all 5% treated specimens. All statistical data were depicted in Table 1. The surface strength of specimens increased gradually from 0% to 3%. However, the strength increased substantially from 3% co-polymer treated specimen (mean = 3772 kPa) to 5% co-polymer treated specimen (17,926 kPa). As previously mentioned, the surface of specimens mixed with 5% co-polymer suppressant was too hard to penetrate through, the actual strength of 5% co-polymer treated soil specimens need further estimation.

3.3 Dynamic Rolling (Grinding) Test

In addition to the above tests, it is interesting to assess how easily fugitive dust emissions can be generated from traffic loading. However, there is no testing standards available that would allow to compare the effectiveness of polymer-treated soil

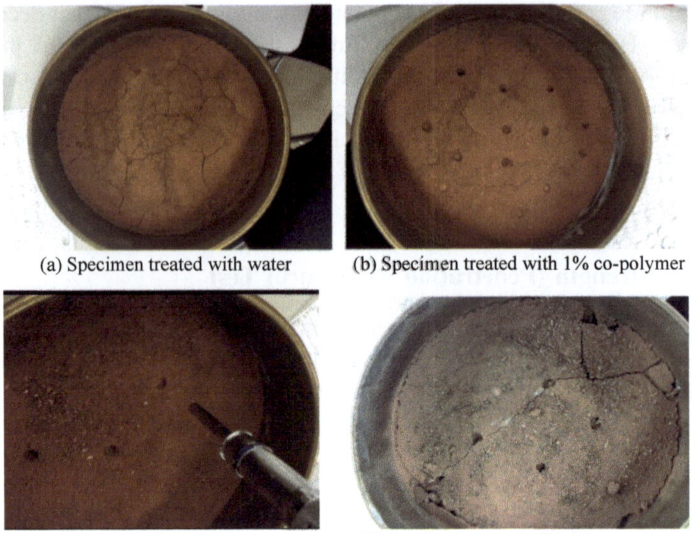

(a) Specimen treated with water (b) Specimen treated with 1% co-polymer

(c) Specimen treated with 3% co-polymer (d) Specimen treated with 5% co-polymer

Fig. 3. Surface conditions of penetrometer tests

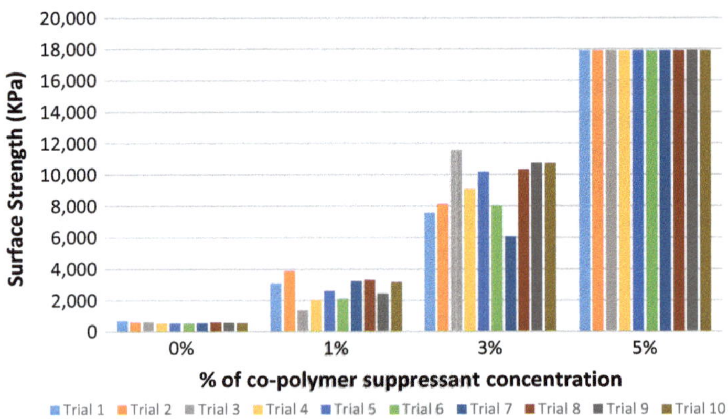

Fig. 4. Surface strength of soil specimens

Table 1. Statistics of surface strength of four specimens

Suppressant Concentration (%)	Ave. (Kpa)	SD (Kpa)	COV
0	771.68	236.2778	32.3%
1	1,949.87	300.4072	16.2%
3	3,772.275	606.0165	16.9%
5	17,914	N/A	N/A

specimens in m mitigating fugitive dust emission. To address the issue, the team search for any testing devices in the laboratory and determined to use the mechanical shaker to facilitate the assessment. Hence, a dynamic rolling (grinding) test was conducted with an intent to simulate the effect of dynamic traffic loading on the performance of co-polymer treated specimens. A total of 12 rubber bunch balls (3.3 cm in diameter, each) were placed on the surface of a soil specimen covered by a 3-kg weight mass (18 cm in diameter), and the entire set was then located on a mechanical shaker running/shaking for 6 min. During shaking, rubber bunch balls were able to roll over the surface of a soil specimen with a means to simulate dynamic vehicle travelling on road surface. After completion of a dynamic rolling test, the remaining soil of each specimen was used to perform a sieve analysis and the particle size distribution curve was created which was used to compare with the original curve (Fig. 5).

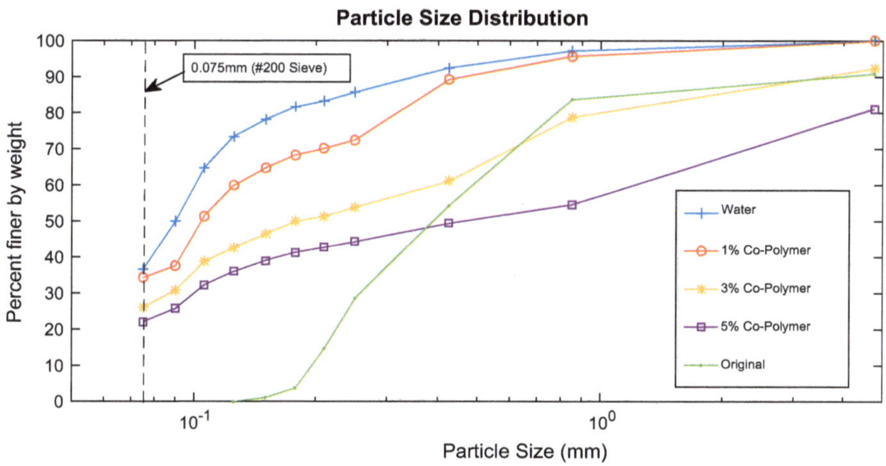

Fig. 5. Soil distribution curves before and after dynamic rolling test

The particles passing #200 sieve is an indication that shows how easily fugitive dust emissions can be generated due to vehicle travelling/traffic loading on road surface. In Fig. 5, the percentage of passing #200 of a specimen drops as the concentration rate of co-polymer treated specimen increases. As shown in Table 2, the soil treated with moisture shows the least resistance (36.7% of particles passing #200) to dynamic rolling (traffic loading) followed by the 1% co-polymer treated specimen (34.4% of particles passing #200), 3% co-polymer treated specimen (26.3% of particles passing #200), and 5% co-polymer treated specimen (22.3% of particles passing #200). The findings in Fig. 5 and Table 2 indicate that soil specimens mixed with 5% co-polymer dust suppressant exhibit excellent abilities to bond soil particles firmly to resist traffic loading so as to reduce the generation of fugitive dust emissions. The above findings help better understand the interlock and bonding effect of the soil particles with respect to the levels of co-polymer application.

Table 2. Soil particle data after dynamic rolling test

Soil specimens with treatment	Water, 0%	1% Co-polymer	3% Co-polymer	5% Co-polymer
% passing #200	36.7	34.4	26.3	22.3

3.4 Scanning Electron Microscopy Experiment

A scanning electron microscopy (SEM) instrument was employed to enable the researcher to thoroughly examine the microstructures of specimens mixed with different co-polymer suppressants. After completion of surface strength tests using a pocket penetrometer, SEM samples were collected from each specimen prior to the surface strength test and dynamic rolling test. The scanning electron microscope scans a focused electron beam over the surface of a co-polymer treated specimen to create an image in which the microstructure of the specimen was able to be examined and measured. The SEM imaging was performed using a Zeiss Supra 40VP SEM located at Northern Arizona University. During SEM imaging, specimens were zoomed to 215x to allow measuring dimensions and observing soil particle interlock and bonding situations with respect to the levels of co-polymer treatment. Figure 6 exhibits SEM images with 215x. it can be seen that soil particles were covered by thin films created by the application of co-polymer product. In comparison with the 1% and 3% co-polymer treated specimens, the file soil particles within the specimen treated with 5% co-polymer tend to adhere to the coarse particles to form a stronger skeleton that would provide higher strength of the soil as evidenced by the surface strength. This demonstrates that as the concentration of the co-polymer increases, the cohesion and interlock effect between particles increases first, and then deceases.

4 Conclusions

This paper presents a lab experiment approach to quantify mechanical responses of co-polymer dust suppressant mixed with native soil (clay soil) in Northern Arizona. Through a series of lab experiments, the following conclusions are drawn:

1. The moisture retention capacity of specimens treated with co-polymer suppressant is better than that of the specimen treated with water. As the concentration of the co-polymer suppressant increases, the moisture retention capacity of the specimens increases as well.
2. Co-polymer treated specimens exhibit promising surface strength than water-treated specimen. The surface strength and resistance to load penetration of specimens increases with the increase of co-polymer suppressant concentration. Surface strength of specimens mixed with 5% co-polymer dust suppressant exceeded the maximum capacity of the penetrometer, so the actual resistance reading cannot be obtained. A further estimate need to be done to obtain actual surface strength of 5% co-polymer treated soil specimens.

(a) Specimen with 1% co-polymer treatment

(b) Specimen with 3% co-polymer treatment

(c) Specimen with 5% co polymer treatment

Fig. 6. SEM 215x images of co-polymer treated specimens

3. Compared with the water-treated specimen, the percentages of soil particles passing through 0.075 mm (sieve #200) made of co-polymer dust suppressants is less than that of the water-treated specimen after dynamic rolling tests. After dynamic rolling on soil specimens, 5% co-polymer treated specimens were able to generate less finer fugitive dust emissions.
4. The SEM images show that higher concentration rate of co-polymer suppressant is able to create thin films to increase the cohesion and interlock effect of the specimens, which contributed to the decrease in generation of fugitive dust during construction, mining, and vehicle travelling on unpaved roads.
5. The use of co-polymer dust suppressant has successfully demonstrated its abilities to improve the mechanical performance of unpaved roads and mitigate the generation of fugitive dust emissions.

Acknowledgements. This research work is supported by the following grants: (1) Natural Science Foundation of Hunan Province, China (18JJ4028), (2) the Foundation of National Key Laboratory for Safety and Health of Mines sponsored by the MaanShan Institute of Mining Research CO., LTD., SinaSteel Group, China (2017-JSKSSYS-05), (3) the Industrial Collaboration Funding of Hengyang Science and Technology Program, China (2015kc58), (4) 2016 Technology Funding for Safe Production and Severe Incident Prevention (Hunan-0003-2016AQ), and (5) 2017 Technology Funding for Safe Production and Severe Incident Prevention (Hunan-0007-2017AQ), and (6) the Opening Foundation Project of Hunan Provincial Key Laboratory of Safe Mining Techniques of Coal Mines at Hunan University of Science and Technology, China (201405). The authors expressed their gratitude for the above funding supports. The authors also would like to thank Desert Mountain Inc. for providing polymer dust suppressants to be used in the experiments. The assistance from staff of Desert Mountain Inc. is gratefully appreciated.

References

Bai, X.B., Liu, J., Yan, I.T.: Fugitive dust pollution of city and situation and prospect of study on dust-depressor. J. Shaanxi Univ. Technol. (2005). (In Chinese)

Bengough, A.G., Mullins, C.E.: Mechanical impedance to root growth: a review of experimental techniques and root growth responses. J. Soil Sci. (1990). https://doi.org/10.1111/j.1365-2389.1990.tb00070.x

Bolander, P., Yamada, A.: Dust Palliative Selection and Application Guide. Project Rep. 9977-1207-SCTDC, U.S. Dept. of Agriculture, Forest Service, San Dimas Technology and Development Center, San Dimas, CA (1999)

Bradford, J.M., Grossman, R.B.: In-situ measurement of nearsurface soil strength by the fall-cone device. Soil Sci. Soc. Am. J. **46**(4), 685–688 (1982)

Campbell, D.J., O'Sullivan, M.F.: The cone penetrometer in relation to trafficability, compaction, and tillage. In: Smith, K., Mullins, C.E. (eds.) Soil Analysis: Physical Methods, pp. 399–429. Marcel Dekker, New York (1990)

Centers for Disease Control and Prevention: Silicosis mortality, prevention, and control—United States, 1968–2002. Morb. Mortal. Wkly Rep. **54**(16), 401–405 (2005)

Chen, R., Lee, I., Zhang L.Y.: Biopolymer stabilization of mine tailings for dust control. J. Geotech. Geoenviron. Eng. (2015). doi: https://doi.org/10.1061/(asce)gt.1943-5606.0001240

Cooper, C.D., Alley, F.C.: Air Pollution Control-A Design Approach, 4th edn. Waveland Press, Incorporated (2011)
EPA: National summary of particulate matter emissions in 2002, Triangle Park, NC. Available: http://www.epa.gov/air/emissions/pm.htm#pmnat (2006). Accessed July 2017
Gillette, D.A., Adams, J., Muhs, D., Kihl, R.: Threshold friction velocities and rupture moduli for crusted desert soils for the input of soil particles into the air. J. Geophys. Res. (1982). https://doi.org/10.1029/JC087iC11p09003
Govers, G., Poesen, J.: Afield-scale study of surface sealing and compaction of loam and sandy loam soils. Part 1. Spatial variability of surface sealing and crusting. In: Callebaut, F., Gabriels, D., DeBoodt, M. (eds.) Proceedings of the Symposium on Assessment of Soil Surface Sealing and Crusting, pp. 171–182. Flanders Research Centre for Soil Erosion and Soil Conservation, Ghent, Belgium (1986)
Midwest Research Institute: Background Document for Revisions to Fine Fraction Ratios Used for AP-42 Fugitive Dust Emission Factors (2006)
Rice, M.A., Mullins, C.E., McEwan, I.K.: An analysis of soil crust strength in relation to potential abrasion by saltating particles. Earth Surf. Process. Land. **22**, 869–883 (1997)
Richards, L.A.: Modulus of rupture as an index of crusting of soil. Soil Sci. Soc. Am. J. **17**(4), 321–323 (1953)
Sanders, T.G., Quayenortey, J.A., Jorgensen, D.: Unpaved road dust control in the Piceance Creek Basin in Rio Blanco County, Colorado. J. Transp. Eng. (2015). doi: https://doi.org/10.1061/(asce)te.1943-5436.0000706
Skidmore, E.L., Powers, D.H.: Dry soil-aggregate stability: energy-based index. Soil Sci. Soc. Am. J. **46**(6), 1274–1279 (1982)
Wang, X., Wang, W.J.: Urban dust pollution and its research of control method in China. Environ. Sci. Technol. **37**(120), 588–592 (2014). (In Chinese)
Xi, Z., Feng, Z., Li, A.: Synergistic coal dust control using aqueous solutions of thermoplastic powder and anionic surfactant. ElsevierB.V., (2017)

A Different Perspective on the Production and Application of Warm Mix Asphalt Under Unfavorable Temperature Conditions

Rui Micaelo[1], José Neves[2(✉)], Rui Lopes[3], and Alexandra Jesus[3]

[1] CERis, Department of Civil Engineering, FCT, Universidade Nova de Lisboa, Caparica, Portugal
[2] CERis, Department of Civil Engineering, Architecture and Georesources, Instituto Superior Técnico, Universidade de Lisboa, Lisbon, Portugal
jose.manuel.neves@tecnico.ulisboa.pt
[3] FCT, Universidade Nova de Lisboa, Caparica, Portugal

Abstract. Warm mix asphalt (WMA) is produced by a variety of technologies at lower temperatures that enable to gain important environmental and social benefits and, in consequence, to contribute to a more sustainable transportation infrastructure. The producer usually defines the temperatures used in WMA production. However, some references suggest the increase of temperature in order to allow a longer time to transport or to compact in cases of unfavourable weather conditions, mainly under low temperatures. The objective of this paper is to analyse the feasibility of producing WMA at the same temperatures of hot mix asphalt (HMA) guaranteeing an adequate compaction, and final performance in service. The paper describes a laboratory study to investigate the properties of a WMA using different additives and varying the mixing and compaction temperatures. The effect of production temperatures on the performance of the WMA was evaluated through binder drainage (production phase), volumetric properties (compaction phase), and resistance to permanent deformation (service phase). The paper also presents a numerical study on the time available for paving WMA under unfavourable climatic conditions. Results demonstrated that it is possible to produce WMA at high temperatures without problems of binder drainage, during transport, and of performance in service if adequate compaction is achieved. In fact, the production temperatures influenced the compaction phase. However, it is possible to increase the temperature without negatively affecting the required volumetric properties. The rut depth of the permanent deformation test was mostly influenced by the air-voids of the compacted WMA and the binder. From the numerical study, it was concluded that the time available for in situ compaction increased substantially when WMA was paved at higher temperature. However, in cases of low air temperature and thin layer, the increase of temperature may not be sufficient to obtain the desired level of density or air-voids.

M. Barman et al. (eds.), *Transportation and Geotechniques: Materials, Sustainability and Climate*, Sustainable Civil Infrastructures,
https://doi.org/10.1007/978-3-319-95768-5_3

1 Introduction

The interest in paving techniques with less environmental and social impacts has increased tremendously in the 21st century. Warm mix asphalt (WMA), which require lower production temperature than traditional hot-mix asphalt (HMA), are the most reported and adopted among different paving techniques (Micaelo et al. 2014a). The temperature reduction during production is about 25 °C and paving is performed with the same methods and techniques. Different methods can be used to lower production temperatures: organic additives; chemical additives; and foamed bitumen.

In the literature (Hurley and Prowell 2005; Hurley and Prowell 2006; Rubio et al. 2011; Jamshidi et al. 2013), the reported economic, environmental and social benefits on the use of WMA are:

- lower energy consumption during asphalt fabrication;
- lower cost of production;
- reduction of emissions and odours;
- better in-site working conditions;
- lower binder aging during production.

However, benefits vary with the method used to lower production temperatures. For instance, when using additives the WMA production cost is higher than of HMA because savings in energy consumption are less than additives cost (Silva et al. 2010a, b). On the other hand, environmental and social benefits are highly valuable in some countries, therefore being adopted in regular paving (EAPA 2014).

In this study three types of additives were used to produce WMA: two organic additives (Sasol Wax 2015; Ventraco Chemie2015); one chemical additive (Akzonobel 2015). In general, these additives have been often used in laboratory studies and in the rehabilitation and construction of roads.

Organic additives are synthetic waxes, obtained with various industrial procedures (Neves et al. 2013). Their melting point varies between 68 °C and 115 °C and fully soluble in bitumen at temperatures above 115 °C. When melted the viscosity of wax is lower than of bitumen. The result is lower viscosity modified bitumen at temperatures above the wax melting point. This structural change allows reducing the temperature of asphalt mixture for mixing and compaction operations. However, the modified binder is comparatively stiffer in the in-service temperature range (Silva et al. 2010a).

Chemical additives are composed by surfactant and adhesion promoter agents (chemical and organic) (Capitão et al. 2012). Surfactants ease aggregates cover by bitumen (reduction of surface tension). Adhesion agents improve active adhesion between materials with no effect on binder viscosity.

In addition, the WMA binder used in the study described in this paper is a specifically designed binder for WMA applications, whose information on additives and modification process was not reported by the supplier.

Operations in paving and compaction are conditioned by the time period available. This time period is a function of the weather conditions (air and ground temperatures, wind speed, clouds cover, solar radiation) and the layer characteristics (thickness, workability and compactibility of asphalt mixture) (Micaelo et al. 2014b). Workability and compactibility are influenced by asphalt mixtures' properties, and the production method in the case of WMA. In most countries, paving is restricted during winter months because the time period available for compaction is insufficient. To solve this, Crews (2008) reported some case-studies, in the USA, wherein WMA mixtures (chemical additives) were fabricated at the traditional temperatures of HMA to extend the time period available for paving. Air temperature was considerably low. Thus, he reported adequate compaction attained in all situations.

However, WMA produced at different temperature may have different performance. Asphalt pavement performance is influenced by asphalt microstructure, which is a function of the components used and the voids structure obtained during compaction. Moreover, the fabrication process affects the bitumen/mastic layer covering the coarse aggregates.

The paper describes a laboratory study with the objective to investigate the properties of a WMA using different additives and varying the mixing and compaction temperatures. The effect of production temperatures on the performance of the WMA was evaluated through laboratory tests: binder drainage (production/transport phase), volumetric properties (compaction phase), and resistance to permanent deformation (in-service performance). The paper also presents a numerical study on the time available for paving WMA under unfavourable climatic conditions.

2 Experimental Research

2.1 Materials

An asphalt concrete AC 20 base 35/50 was used in this study. This mixture was designed without WMA additives in accordance to EN 13108-1 (Martinho 2014). The binder is a paving grade bitumen, penetration grade 35/50 (10^{-1} mm) (see Table 1). Table 2 lists the main physical properties of the asphalt mixture. Figure 1 presents the grading curve of the aggregates, including the standard limits.

Table 1. Characteristics of bitumen

Property	Standard	35/50	BT 35/50
Penetration @ 25 °C [0,1 mm]	EN 1426	47	35–50
Softening point [°C]	EN 1427	51.4	50–58
Durability			
Retained penetration [%]	EN 1426	≥ 53	≥ 53
Increase softening point [°C]	EN 1427	≤ 11	≤ 11
Change of mass [%]	EN 12607-1	≤ 0.5	≤ 0.5
Fraass point [°C]	EN 12593	≤ -5	≤ -5
Flash point [°C]	EN ISO 2592	≥ 240	≥ 240

The phases of mixing and compaction of the asphalt mixture were performed at several combinations of temperatures to investigate their effect on the production and

Table 2. Asphalt mixture design properties (Martinho 2014)

Property	Unit	AC 20 base 35/50
Binder content	$\%_{Mass}$	4.5
Bulk density	kg/m^3	2360
Max density	kg/m^3	2455
Air voids	%	3.9
Voids in mineral aggregate	%	14.8
Voids filled with bitumen	%	69.6

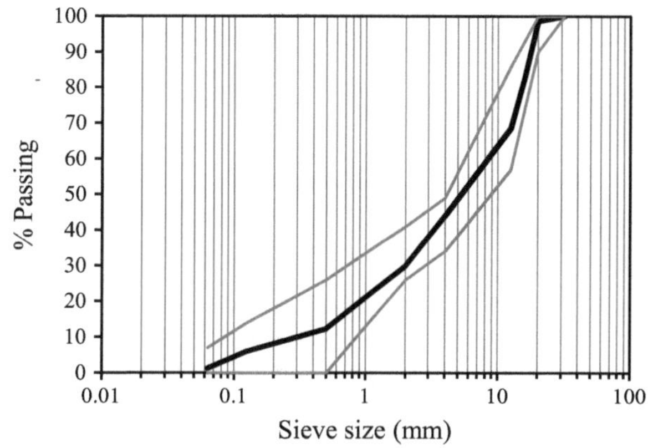

Fig. 1. Grading curve of the aggregate – AC20 base 35/50

performance of WMA. Standard production temperatures of AC 20 base 35/50 were 165 °C (mixing) and 150 °C (compaction). To reduce production temperatures of this mixture three different additives – AdA, AdB, and AdQ – mixed in laboratory and one WMA modified binder by the producer in origin – BT – were used (see Table 1). Additives AdA and AdB are organic and additive AdQ is chemical. The content of additive added to bitumen 35/50 was 4% and 2% (w/w) for the organic and chemical additives, respectively. The producer did not disclose additives used in BT. Figure 2 plots the dynamic viscosity versus the temperature (120–165 °C) for paving grade bitumen and WMA modified binders. Dynamic viscosity measurements were performed with a rotational viscometer (Selecta ST-2020 R), using a T-bar (R2) type spindle. The largest reduction in viscosity occurred for organic additive AdB. Unexpectedly, chemical additive AdQ and organic additive AdA had similar effect (reduction) on viscosity. The least difference in viscosity to paving grade bitumen was showed by BT.

Table 3 lists temperatures used for mixing and compaction of different asphalt mixtures. Asphalt mixture with paving grade bitumen was produced at reference temperatures (165 °C and 150 °C). Production temperatures of WMA vary with the type and amount of additive used, but are commonly near 120 °C (Martinho 2014). Hence, mixing and compaction at 120 °C is the reference for WMA. In addition, these

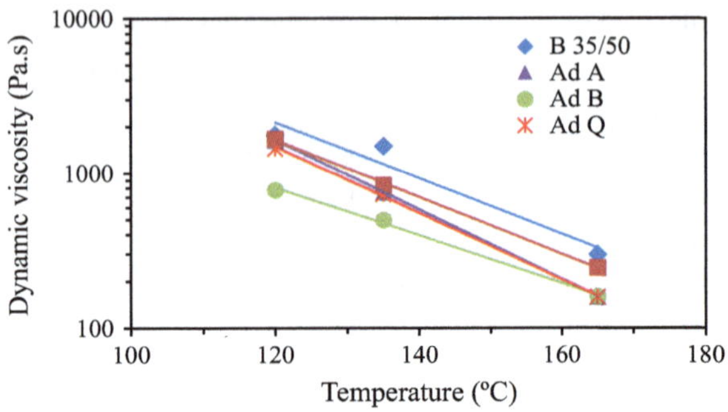

Fig. 2. Variation of dynamic viscosity with temperature

Table 3. Mixing and compaction temperatures

Mixing	Compaction	Binder	Reference
165 °C	150 °C	35/50	B - F165/C150
	120 °C	35/50 + 4% organic additive AdA	AdA - F165/C120
		35/50 + 4% organic additive AdB	AdB - F165/C120
		35/50 + 2% chemical additive AdQ	AdQ - F165/C120
		35/50 modified	BT - F165/C120
135 °C	120 °C	35/50 modified	BT - F135/C120
120 °C	120/100 °C	35/50 + 4% organic additive AdA	AdA - F120/C120 AdA - F120/C100
		35/50 + 4% organic additive AdB	AdB - F120/C120 AdB - F120/C100
		35/50 + 2% chemical additive AdQ	AdQ - F120/C120 AdQ - F120/C100
		35/50 modified	BT - F120/C120 BT - F120/C100

mixtures were produced at 165 °C and compacted at 120 °C, and produced at 120 °C and compacted at 100 °C. Thus, temperatures recommended by the producer of BT (135 °C and 120 °C) were also included in the study. Before mixing, aggregates were heated at the mixing temperature plus 20 °C for 4 h to ensure drying of aggregates.

The cylindrical test specimens used for the water sensitivity analysis were compacted using an impact compactor, with 75 blows applied in each side of the specimen, according to EN 12697-30. The cylindrical specimens (300 × 400 × 60 mm) used in the wheel-tracking tests (rutting resistance) were compacted by a laboratory roller compactor, with controlled compaction energy setting, in accordance to EN 12697-33.

2.2 Laboratory Tests

The influence of mixing temperatures on production, paving and in-service performance of WMA was evaluated based on three different tests: binder drainage, water sensitivity and rutting resistance.

Binder drainage testing was performed to evaluate if the increase in mixing temperature causes segregation in mixture due to binder and mastic drainage during transport. Testing was performed with WMA mixtures produced at 165 °C (highest temperature), using the basket method in accordance to EN 12697-18. Briefly, the test method consists in loading a perforated basket (see Fig. 3) and placing it in an oven at test temperature for approximately 3 h. Then, the metal tray used under the basket is weighed to determine the mass of binder and mastic that drained from the basket. Drainage (D) is the ratio of drained mass to initial mass sample in the basket.

Fig. 3. Steel basket used for binder drainage testing

Water sensitivity analysis aims at determining the resistance of aggregate to mastic bond to the water damaging effect. Testing was performed in accordance to EN 12697-12. Two similar sets of cylindrical specimens (3) were conditioned in different environments, one dry at 20 °C (dry group) and the other in water at 40 °C (wet group). Then, the specimens were tested under indirect tensile testing at 15 °C. The ratio (ITSR) of the wet (ITSw) to dry (ITSd) indirect tensile strength average values was used to evaluate the water sensitivity of the mixture.

The wheel-tracking test, performed in accordance to EN 12697-22, was used to evaluate the resistance of asphalt mixture to deform under repeated loading. Prismatic specimens were trafficked (continuous back and forward) at elevated temperature (60 °C) by a constant load (600 kPa) wheel, and the rut depth on the centre of the specimen was measured in each wheel pass. Two specimens were tested for each asphalt mixture. Resistance to permanent deformation was determined from the final rut depth (RD_{AIR}, mm) and the wheel-tracking slope (WTS_{AIR}, mm/1000 cycles) calculated as the average rate at which the rut depth increases with repeated passes of a loaded wheel:

$$WTS_{AIR} = \frac{d_{10000} - d_{5000}}{5}$$

where d_{10000} and d_{5000} are the rut depth (mm) after 10000 and 5000 loading cycles, respectively.

3 Results and Discussion

Drainage value (D) for different WMA mixtures was zero, which means these asphalt mixtures can be stored (or transported) at high temperature (165 °C) without segregation issues. Hence, drainage would not be expected for a dense graded asphalt mixture with hard paving bitumen. However, additives segregation issues in WMA are conceivable. Chemical additives are packages of surfactant and adhesion agents that affect bonding and aggregate cover by the bitumen, whereas organic additives are usually waxes that reduce the viscosity of binder at temperatures above the wax melting point (around 100 °C). Although these mixtures were mixed and stored at a temperature 40 °C higher than recommended by additives suppliers there was not binder drainage.

Figure 4 presents average air voids of compacted cylindrical specimens. In general, air voids of specimens were higher than of the reference mixture. Mixing and compaction temperatures had a clear influence on the final air voids but there was not a clear defined trend. For instance, air voids increased with decreasing temperatures for mixtures with additive AdA whereas the opposite occurred for mixtures with additive AdB. For BT, as-designed air voids were obtained with the set of temperatures recommended by the supplier for this binder (135 °C/120 °C).

Water sensitivity test results are presented in Fig. 5. Most of asphalt mixtures showed large resistance to water damage. Only four mixtures had an ITSR value lower than 80%. With exception to additive AdA, there was a trend of increased water sensitivity with decreasing production temperatures.

Wheel-tracking test results are listed in Table 4. Air voids of the prismatic specimens are also included in this table. In general, air voids (4.7–11.5%) were significantly higher than the asphalt mixture design value (3.9%). The lowest values were obtained for the BT. Thus, air voids values were more affected by the type of additive used than by production temperatures (1–2%). The low compaction level needs to be considered in the analysis.

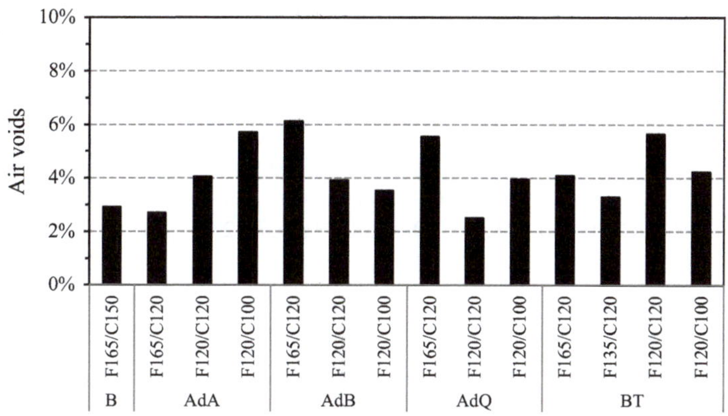

Fig. 4. Variation of air voids with mixing and compaction temperatures

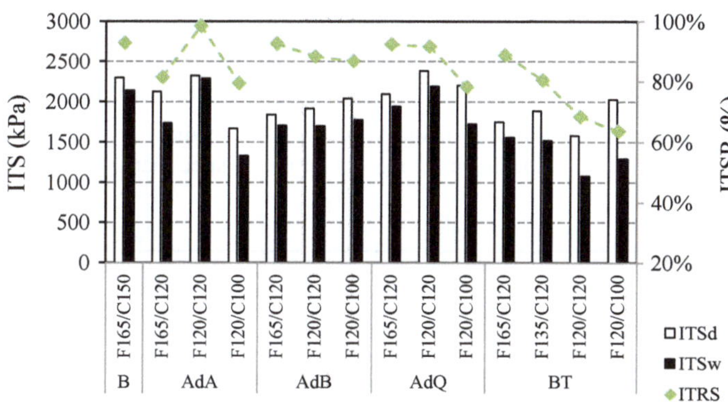

Fig. 5. Water sensitivity test results

Figure 6 shows the evolution of ruth-depth during the wheel-tracking tests. The performance under repeated loading of reference specimen (B) was as expected for a dense grade asphalt mixture with a hard penetration grade (non-modified bitumen). The rut depth, after 20000 cycles, was about 5 mm, and the wheel-tracking slope was 0.272 mm/1000 cycles. Most of tests with additives AdA and AdQ were interrupted before the 20000 cycles because maximum rut depth was attained. In addition, specimens made with additive AdB had lower resistance to permanent deformation than reference. The opposite was concluded for specimens with BT.

In literature (Neves et al. 2013; Silva et al. 2010b) is referred that waxes (organic additives) reduce the viscosity of the bitumen above the wax melting point (see Fig. 2), and the opposite occurs at lower temperatures (in-service temperatures). Hence, it was expected that specimens with organic additives (AdA and AdB) had a strong rutting resistance. In case asphalt mixtures with additive AdA, specimens were actually better compacted than of reference mixture.

Table 4. Wheel-tracking test results

Reference		Air voids (%)			WTS_{AIR} (mm/10^3ciclos)			RD_{AIR} (mm)		
		#1	#2	Avg	#1	#2	Avg	#1	#2	Avg
B	F165/C150	9.2	9.0	9.1	N.v.	0.272	0.272	n.v.	4.77	4.77
AdA	F165/C120	6.4	6.8	6.6	2.863	n.v.	2.863	20[a]	20[a]	20
	F120/C120	7.7	8.0	7.9	2.831	n.v.	2.831	20[a]	20[a]	20
	F120/C100	7.5	6.8	7.1	n.v.	n.v.	n.v.	20[a]	20[a]	20
AdB	F165/C120	8.0	8.9	8.4	0.172	0.173	0.173	4.40	5.44	4.92
	F120/C120	8.7	10.0	9.4	0.316	0.412	0.364	6.66	7.02	6.84
	F120/C100	8.7	9.6	9.2	0.844	0.166	0.505	5.44	3.32	4.38
AdQ	F165/C120	9.5	8.3	8.9	1.943	n.v.	1.943	20[a]	n.v.	20
	F120/C120	11.5	10.4	11.0	1.548	n.v.	1.548	20[a]	20[a]	20
	F120/C100	9.6	9.1	9.4	1.815	0.786	1.301	20[a]	13.34	16.67
BT	F165/C120	5.8	5.4	5.6	0.088	0.080	0.084	3.95	2.97	3.46
	F135/C120	5.7	5.7	5.7	0.082	0.100	0.091	3.62	3.75	3.69
	F120/C120	6.7	6.1	6.4	0.124	0.120	0.122	3.72	3.68	3.70
	F120/C100	4.7	5.8	5.3	0.264	0.256	0.260	6.47	6.07	6.27

[a]Stopped before 10,000 loading cycles; n.v. – test not valid.
#1 – specimen 1; #2 – specimen 2; avg – average of specimens 1 and 2 results

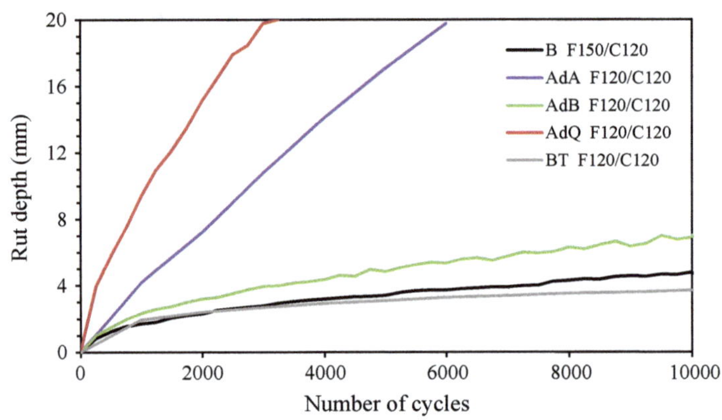

Fig. 6. Evolution of rut depth in wheel-tracking tests

Moreover, the effect of production temperatures on rutting resistance can only be evaluated for additive AdB and BT. For both materials there was one set of temperatures with lesser performance: temperatures (mixing/compaction) were 120 °C/120 °C for additive AdB, and 120 °C/100 °C for BT. However, these specimens were not the least compacted. Hence, as concluded in other studies (Micaelo et al. 2016), distribution of aggregate particles and bitumen in compacted specimens changes with production temperatures and is not captured by global volumetric indexes, such as air voids.

Diverse distribution patterns were possible for similar air voids, which affects in-service performance.

These results show that WMA can be produced at the same temperatures commonly used with HMA, without segregation issues during transport and similar in-service performance. However, adequate field compaction has to be achieved. In general, WMA produced at high temperatures had similar final air voids values to the specimens mixed and compacted at the recommended temperatures.

4 Time Available for Compaction

The experimental results presented and discussed in previous section showed that WMA layers can be effectively built with mixing performed at higher than recommended temperatures. Hence, this section is aimed at determining the increase in available paving time in cold weather conditions for WMA mixtures produced at typical HMA temperatures. To this analysis, the analytical model presented in Alves (2015) to estimate the temperature of asphalt layer was used. Temperature evolution was estimated by means of a one-dimensional heat flux model, considering heat transfer by conduction, convection and radiation. Table 5 presents four case-studies analysed in the paper. In the literature there are not specific values of thermal properties for WMA binders and mixtures. Hence, material properties used in the model (see Table 5) were based on values used by other researchers for HMA. Weather and paving conditions considered in the analysis are presented in Table 6.

Table 5. Time available for compaction – structures

Layers	Case study			
	1	2	3	4
Paved	Asphalt concrete h = 30 mm λ = 2.0 W/m. °C c = 869 J/kg. °C ρ = 2350 kg/m^3	SMA h = 50 mm λ = 1.2 W/m. °C c = 921 J/kg. °C ρ = 2400 kg/m^3	Asphalt concrete h = 80 mm λ = 2.0 W/m. °C c = 869 J/kg. °C ρ = 2350 kg/m^3	Asphalt concrete h = 120 mm λ = 2.0 W/m. °C c = 869 J/kg. °C ρ = 2350 kg/m^3
Base	Asphalt concrete h = 200 mm λ = 2.0 W/m. °C c = 869 J/kg. °C ρ = 2350 kg/m^3	Asphalt concrete h = 200 mm λ = 2.0 W/m. °C c = 869 J/kg. °C ρ = 2350 kg/m^3	Granular h = 200 mm λ = 1.3 W/m. °C c = 837 J/kg. °C ρ = 1750 kg/m^3	Granular h = 200 mm λ = 1.3 W/m. °C c = 837 J/kg. °C ρ = 1750 kg/m^3

Figure 7 plots the evolution of layer temperature with time for the four structures, in three different weather situations (air temperature). Pavement layer temperature corresponds to average temperature in all determined points (10 mm interval). Variation of the layer thickness during cooling was not considered. Often, 6 to 12 roller passes (one direction) were needed to attain desired compaction level depending of compaction conditions (asphalt mixture, layer thickness, available rollers). This means that at regular roller operation speed, at least 5 min are needed to perform compaction.

Table 6. Time available for compaction – weather conditions

Paving temperature (°C)	120	160	
Air temperature (°C)	0	5	10
Base temperature (°C)	3	9	14
Wind speed (m/s)	6 (moderate)		
Sky	Clear		

In the analysis (red coloured line in Fig. 7) was admitted that compaction cannot be adequately performed at temperatures below 100 °C. In this situation, thin to medium layers (30–50 mm, case-studies 1 and 2) cannot be adequately paved, even if asphalt mixtures are produced at 160 °C.

For case-studies 3 and 5, thicker paving layers, there was a significant increase in time available for compaction when production temperature is 160 °C. In case of regular production temperature (120 °C), compaction was not likely to be attained at higher weather temperatures (10 °C), whereas producing at 160 °C allowed achieving compaction with the lowest air temperature (0 °C).

5 Conclusions

This paper presented and discussed the results of a study aimed at assessing the influence of WMA production temperatures in paving and in-service performance. Some references suggest the increase of production temperature in order to allow a longer time to transport or to compact in cases of unfavourable weather conditions, mainly under low temperatures.

A standard dense grade asphalt mixture (AC 20 base 35/50) was produced at various temperatures. To allow reducing production temperatures, three different WMA additives, mixed with the paving grade bitumen modified, and one WMA specific binder were used. WMA mixtures were produced at 165 °C, 135 °C and 120 °C, and compacted at 120 °C and 100 °C. The effect of temperatures was evaluated based on four indicators: binder drainage (production/transport phase); volumetric properties of compacted specimens (compaction phase); and water sensitivity and rutting resistance (in-service performance). Additionally, it was analysed the effect of production temperatures on the estimated time available for compaction under low air temperature conditions.

Binder drainage did not occur with any of WMA mixtures produced at 165 °C. Despite the general problems in achieving desired compaction, compaction was not worsened by increasing production temperatures (HMA temperatures). Actually, in some cases compaction was better. Rutting resistance results were significantly affected by final air voids of compacted specimens and binder used (with and without additives).

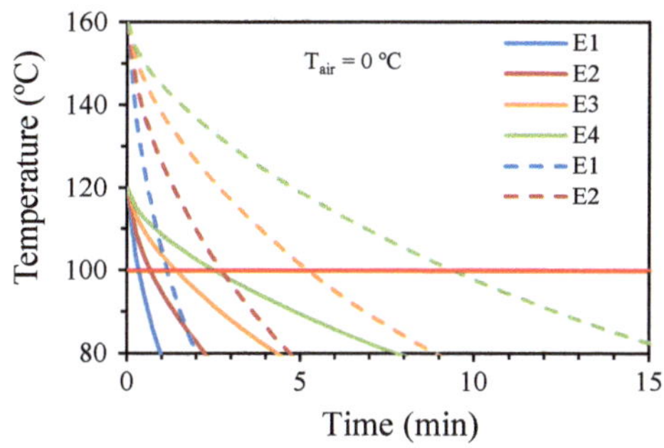

(a) Air temperature: T_{air} = 0 °C

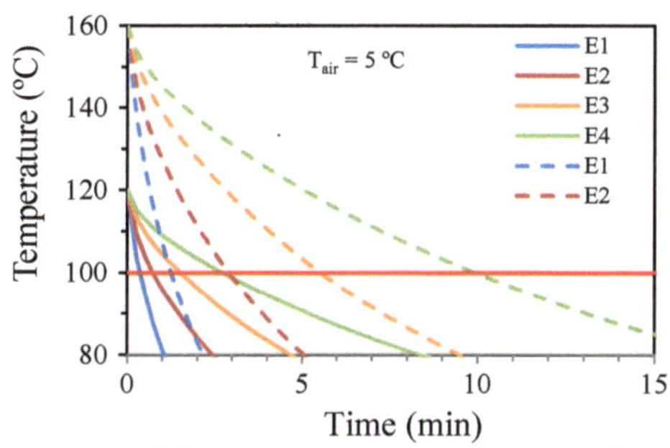

(b) Air temperature: T_{air} = 5 °C

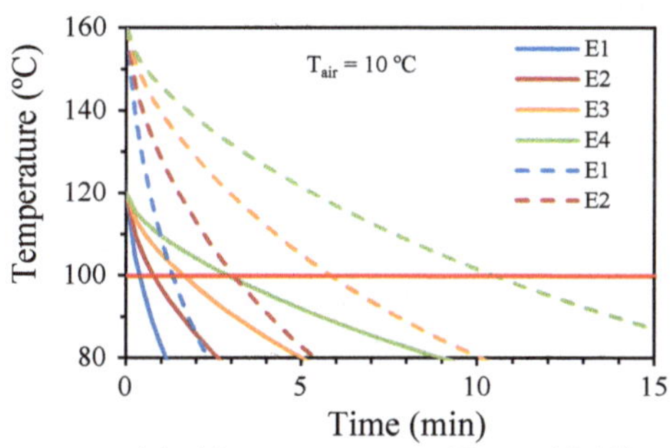

(c) Air temperature: T_{air} = 10 °C

Fig. 7. Evolution of internal temperature with time

The time period for in-field roller compaction was extended significantly when asphalt mixture was paved at higher temperatures. Nevertheless, in situations of low air temperature and thin paved layer, the increase in asphalt temperature was not sufficient to allow achieving desired compaction.

References

Alves, M.R.: Time period available asphalt compaction: parametric analysis (in Portuguese). M. Sc. thesis, Universidade Nova de Lisboa, Portugal (2015)

Akzonobel Website: https://www.akzonobel.com/innovation/our_innovations/rediset/ [visited in July 2015]

Capitão, S., et al.: Pavement engineering materials: review on the use of warm-mix asphalt. Constr. Build. Mater. **36**, 1016–1024 (2012). https://doi.org/10.1016/j.conbuildmat.2012.06.038

Crews, E.: Extended season paving in New York city using Evotherm™ warm mix asphalt. MeadWestvaco Asphalt Innovations, USA (2008)

EAPA: Asphalt in figures 2013. European Asphalt Pavement Association, Brussels (2014)

Hurley, G.C., Prowell, B.D.: Evaluation of Sasobit® for use in Warm Mix Asphalt. NCAT Report 05-06. National Center for Asphalt Technology, Auburn University, Auburn (2005)

Hurley, G.C., Prowell, B.D.: Evaluation of Evotherm® for use in Warm Asphalt Mixes. NCAT Report 06-02. National Center for Asphalt Technology, Auburn University, Auburn (2006)

Jamshidi, A., et al.: Performance of Warm Mix Asphalt containing Sasobi: state-of-the-art. Constr. Build. Mater. **38**, 30–553 (2013). https://doi.org/10.1016/j.conbuildmat.2012.08.015

Martinho, F.: Warm mix asphalt with industrial wastes—characterization and technological validation (in Portuguese). Ph.D. thesis, Universidade de Lisboa, Portugal (2014)

Micaelo, R., et al.: Hot-mix asphalt compaction evaluation with field tests. The Baltic J. Road and Bridge Eng. **9**(4), 306–316 (2014). https://doi.org/10.3846/bjrbe.2014.37

Micaelo, R., et al.: TRACC-EXPERT: Tool for the Selection of Paving Techniques Adapted to Climate Change. Sustainability, Eco-efficiency, and Conservation in Transportation Infrastructure Asset Management (2014b)

Micaelo, R., et al.: Production of Hot-Mix Asphalt with PMB: compactability and mechanical behaviour characterization. In: F. Canestrarim, Partl, M.N. (eds.) Proceedings of the 8th RILEM International Symposium on Testing and Characterization of Sustainable and Innovative Bituminous Materials, pp. 231–242. Springer (2016) https://doi.org/10.1007/978-94-017-7342-3

Neves, J. et al.: Experimental study related to the performance of Warm Mix Asphalts (in Portuguese). In: Proceedings of the 7th Portuguese Road Conference, Lisboa, Portugal (2013)

Rubio, et al.: Warm mix asphalt: an overview. J. Clean. Prod. **24**, 76–84 (2011). https://doi.org/10.1016/j.jclepro.2011.11.053

Sasol Wax. Website: http://www.sasolwax.us.com [visited in July 2015]

Silva, H., et al.: Assessment of the performance of warm mix asphalts in road pavements. Int. J. Pavement Res. Technol. **3**(3), 119–127 (2010a)

Silva, H., et al.: Optimization of warm mix asphalts using different blends of binders and synthetic paraffin wax contents. Constr. Build. Mater. **24**(9), 1621–1631 (2010b). https://doi.org/10.1016/j.conbuildmat.2010.02.030

Ventraco Chemie B.V. Website: http://www.ventraco.nl [visited in March 2015]

Effect of Cylinder Size on the Compressive Strength of Concrete CO_2 Curing

Ming-Gin Lee[1(✉)], Yung-Chih Wang[2], Yu-Min SU[3], Yu-Cheng Kan[1], and Shih-Hsuan Kao[2]

[1] Department of Construction Engineering, Chaoyang University of Technology, Taichung 413, Taiwan, R.O.C
{mglee, yckan}@cyut.edu.tw

[2] Department of Civil Engineering, National Central University, Chungli 320, Taiwan, R.O.C
wangyc@ncu.edu.tw, dzxcvbmike@gmail.com

[3] Department of Civil Engineering, National Kaohsiung University of Applied Sciences, Kaohsiung City, Taiwan, R.O.C
yuminsu@kuas.edu.tw

Abstract. This study was conducted to assess the effect of cylinder size on concrete compressive strength by CO_2 curing. The factors studied to evaluate the effect of cylinder size on concrete compressive strength were cylinder size, CO_2 concentration, curing pressure, curing time, and age of specimen at testing. Three different cylinder sizes (5 × 10 cm, 10 × 20 cm, 15 × 30 cm), three CO_2 concentration (50%, 75%, 100%), three curing time (1, 3, 6 h), three curing pressures (2, 4, 8 bars), and four testing ages (3, 7, 28, 90 days) were used in this investigation. A batch size of 0.18 m^3 concrete was used to produce twenty 5 × 10 cm cylinders, twenty 10 × 20 cm cylinders, and twelve 15 × 30 cm cylinders. However, it was necessary that the specimens reach an initial set to demold before the CO_2 curing process was used because the dry CO_2 gas-stream would gradually take the free water away from the specimens. After the CO_2 curing duration, the concrete samples were assessed through the compressive strength, the pressure drop and the temperature expected to go down and rise in the chamber. In general, high early and late compressive strength of the CO_2-cured cylinders were obtained while its strength was mostly higher than the strength of normally water-cured ones. The k_{s5} value obtained from 5 × 10 cm, and 15 × 30 cm cylinders by CO_2 curing ranges between 0.95 and 1.30, this is partly due to the fact that high early compressive strength of the CO_2-cured cylinders was obtained while its early strength was about equal to the 28-day strength of normally water-cured ones. Therefore, special care is still needed to compare the properties of concrete from the two different cylinder sizes by CO_2 curing.

1 Introduction

Accelerated curing is any method by which high early age strength is achieved in concrete. These techniques are especially useful in the prefabrication industry, wherein high early age strength enables the removal of the formwork within 24 hours,

M. Barman et al. (eds.), *Transportation and Geotechniques: Materials, Sustainability and Climate*, Sustainable Civil Infrastructures,
https://doi.org/10.1007/978-3-319-95768-5_4

thereby reducing the cycle time, resulting in cost-saving benefits [1]. Thermal acceleration method in concrete can benefit several operations in the construction industry such as precast concrete fabrication. The quality of a steam cured precast concrete is affected by curing time and temperature variables [2]. Steam curing is advantageous where early strength gain in concrete is important or where additional heat is required to accomplish hydration, as in cold weather. Two methods of steam curing are used: live steam at atmospheric pressure (for enclosed cast-in-place structures and large precast concrete units) and high-pressure steam in autoclaves (for small manufactured units). Only live steam at atmospheric pressure will be discussed here. A typical steam-curing cycle consists of (1) an initial delay prior to steaming, (2) a period for increasing the temperature, (3) a period for holding the maximum temperature constant, and (4) a period for decreasing the temperature [3].

Steam curing might cause micro-cracks along the interface of aggregate and cement paste. It has been reported that the strength of concrete by steam curing could decrease in later stages such as 28 days or 90 days, while compared with concrete cured in air or in water [4]. Additionally, it requires a relatively long period for curing since heat must diffuse inward from the surface and the inherently non-uniform temperature can generate thermal cracking.

One potential technology in Carbon dioxide curing is the use of concrete products as carbon sink through the early age curing. The cement compounds C_3S and C_2S are instantaneously carbonized into calcium carbonate and silica gel, once cement is mixed with water and exposed to the CO_2 gas. If it works, concrete production lines can be set next to the power plants or cement kilns to produce the concrete products using the captured CO_2 as curing agent. CO_2 curing is not a new process; its origins can be traced back hundreds or thousands of years [5]. Humans have used alkaline earth hydroxide cement and mortar as a binder material to construct structures, which harden due to their reaction with CO_2 in the atmosphere. Because of the low concentration of CO_2 in atmosphere and low pressure of CO_2, the diffusion of CO_2 into cement mortar is very slow. This results in a slow strength development of the cement mortar. Since the 1970s, research was conducted in an attempt to understand the carbonation mechanisms and their applications in fast curing of cement and concrete products [5].

The CO_2 curing techniques were also utilized to enhance the properties of recycled concrete aggregates [6, 7]. The results showed that accelerated carbonation can densify recycled aggregates derived from crushed mortars [7]. There was a reduction in water absorption and an increase in apparent density of the recycled aggregates due to the carbonation reactions of the calcium hydroxide and calcium silicate hydrates in the cement mortar [8]. Considering the potential of CO_2 sequestration, the feasibility of using CO_2 curing for concrete blocks prepared with recycled aggregates were studied [9]. The results showed that a short period of CO_2 curing allowed the concrete blocks prepared with recycled aggregate to attain a higher compressive strength than that of the blocks prepared with natural aggregates. Another study showed that besides the water content in the concrete blocks, there are several other factors influencing the CO_2 curing progress, including the gas pressure and curing time [10]. This study further investigated the effect of cylinder size on concrete compressive strength and modulus of elasticity by CO_2 curing. "There is strong evidence that if one uses 10×20 cm cylinder plastic or steel molds, the strength obtained in the 2,900 to 14,500 psi range is expected to be 5% greater than that obtained using 15 × 30 cm cylinder molds. In the lower strength ranges, 2,900 to 8,700 psi, for example, it may be acceptable to assume from a practical perspective that strengths using 10 × 20 cm and 15 × 30 cm molds are equivalent; justification for such an assumption must be determined by standards authorities" [11]. However, some believe that the magnitude of difference in standard deviations is great enough to require twice the number of 10 × 20 cm as 15 × 30 cm cylinders to keep an equal degree of precision [12].

2 Experiments

Preparation of concrete materials is first described, followed by the concrete mixture preparation, factor combinations and carbon dioxide curing test.

2.1 Concrete Materials and Mixtures

The materials used in the experiments were: Type I Portland cement conforming to ASTM C 150 manufactured by the Taiwan cement company, coarse aggregates (crushed stone and crushed gravel), fine aggregate from river sand and water. Table 1 show the mix design of normal concrete used in this study. The normal concrete had a water-to-cement ratio of 0.68 and has a 15 ± 2.5 cm slump. A batch size of 0.18 m^3 was established based on the practical mixing capacity of a 0.3 m^3 concrete mixer. This would produce twenty 5 × 10 cm cylinders, twenty 10 × 20 cm cylinders, and twelve 15 × 30 cm cylinders, and 15% extra volume for waste.

Table 1. Mix Design for Normal Concrete (kg/m^3)

Mix	W/C	Cement	Aggregate	Sand	Water
No.1	0.68	302	885	1021	205

2.2 Concrete Samples

Concrete cylinder specimens with three different cylinder sizes (a diameter of 5 cm and a height of 10 cm, a diameter of 10 cm and a height of 20 cm, a diameter of 15 cm and a height of 30 cm) were prepared according to ASTM C39/39 M-16 [13] and used to investigate the compressive strength development of concrete by CO_2 curing. Two cylinder sizes (a diameter of 5 cm and a height of 10 cm, a diameter of 10 cm and a height of 20 cm) were used to study the elasticity modulus of concrete by CO_2 curing. The objective of this research was to assess the effect of cylinder size on the compressive strength of 5 × 10 cm cylinders, 10 × 20 cm cylinders, and 15 × 30 cm cylinders made from the same batch of concrete by CO_2 curing. Based on the literature review, it was determined that the main variables that could potentially affect the strength ratio were strength range, aggregate type, curing method, and testing age. Therefore, both curing method and testing age were selected to be varied within the experimental phase of this research. The compressive strengths in 3, 7, 28 and 90 days of concretes were selected for quality assurance and quality control testing.

2.3 Carbon Dioxide Pressure Test

The test setup for carbon dioxide curing of concrete included carbon dioxide gas tank, mixed air tank, pressure curing chamber, pressure gauge, vacuum pump, heater, thermometer, regulator, and safety valve. Figure 1a is the photo of the CO_2 curing test setup and Fig. 1b is concrete samples in curing chamber. The carbon dioxide curing system had both mixed air and CO_2 gas tanks so that CO_2 could be diluted to simulate the exhaust gas from the thermal power plant or the cement plant. One-time supply method was initially controlled by a gas regulator and kept at 2, 4, 8 bar to supply gas into the curing chamber.

(a) Carbon Dioxide Curing Test Setup

(b) Concrete Samples in Curing Chamber

Fig. 1. Carbon dioxide curing test setup and concrete samples in curing chamber

Two curing methods, CO_2 curing and water curing were employed in this study. Figure 1 illustrates the CO_2 curing set-up. An airtight steel vessel with a three dimensional (3D) cube 60 × 60 × 60 cm about 216 litters was used as the CO_2 curing chamber, which was vacuumed to −0.1 bar before the CO_2 gas injection. The CO_2 pressure in the chamber was controlled by a gas regulator and kept at 2, 4, 8 bar for 1, 3 and 6 hours curing duration, respectively. The CO_2 curing chamber was placed in a room maintained at 25 ± 1 °C. There were twenty 5 × 10 cm cylinders, twenty 10 × 20 cm cylinders, and twelve 15 × 30 cm cylinders cured in CO_2 curing chamber each time. The factor combinations of carbon dioxide curing test used are shown in Table 2.

Table 2. Factor combinations of carbon dioxide curing test

W/C	Pressure (bar)	CO_2 (%)	Time (hour)
0.68	2	50	1
	4	75	3
	8	100	6

2.4 Notation

A specific specimen identification system was developed to keep track of the data in this study. A cylinder's identity was named in the order of strength, CO_2 concentration, curing duration, and curing pressure. For example, a cylinder with the identity L-100-6-4 would be a cylinder made from the general strength of concrete with 100% CO_2 concentration curing 6 hours under 4-bar pressure and CL represents control sample with normally water-cured.

Throughout this study, the following notation was used:

$$f_{c5} = k_{s5} \times f_{c15} \text{ or } f_{c10} = k_{s10} \times f_{c15} \quad (1)$$

where,

f_{c5} = compressive strength of a 5 × 10 cm cylinder,
f_{c10} = compressive strength of a 10 × 20 cm cylinder,
f_{c15} = compressive strength of a 15 × 30 cm cylinder, andk_{s5} and
k_{s10} = the strength conversion factor, correlating the smaller cylinder to the 15 × 30 cm cylinder strength.

3 Experimental Results and Discussion

3.1 Reaction of Curing Concrete with CO_2

The results of CO_2 pressure and temperature reaction of curing concrete with CO_2 for one hour are displayed in Fig. 2. Figure 2a gave a relationship between curing duration and CO_2 pressure loss. CO_2 pressure decreases as the curing duration increases. Figure 2b shows the plots of reaction temperature versus the curing duration for seven CO_2-cured concretes. It can be seen that the L-100-1-4 CO_2-cured concrete has slightly higher reaction temperature as compared to others due to high pressure drop. The performance of the CO_2 cured cylinders was found through the measurement of pressure drop, the temperature expected to go down and rise, and CO2 was being absorbed by concrete.

3.2 Concrete Compressive Strength Associated with CO_2 Curing

In general, higher CO_2 concentration, longer carbonation time, higher CO_2 pressure could increase the compressive strength of the mortar [14]. Figure 3 shows the effect of CO_2 concentration and curing variables on compressive strength for 15 × 30 cm cylinder. Based on 90-day compressive strength at 100% CO_2 concentration, Fig. 3a, the order of strength from highest to lowest, is L-100-6-2, L-100-1-2, L-100-3-4, L-100-3-2, CL and L-100-1-4, but it may vary for different CO_2 concentrations. It can be seen that the CO_2-cured samples have slightly higher compressive strength as compared with the strength of normally water-cured ones (CL) at curing times ranging from 3 to 90 days. It showed that the strength of concrete increases with an increase in

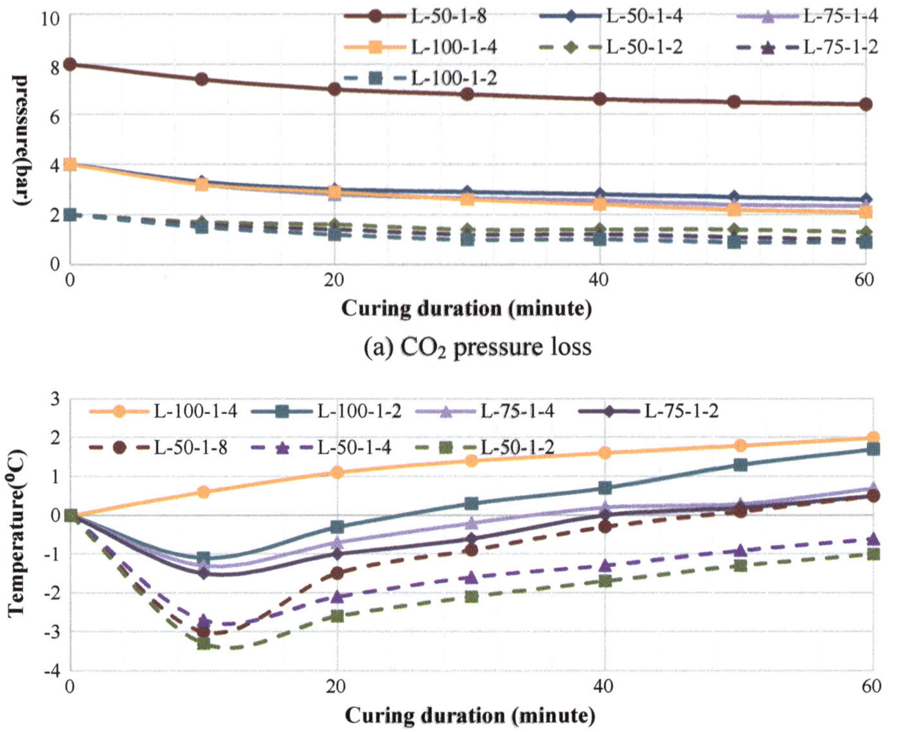

(a) CO_2 pressure loss

(b) temperature rise

Fig. 2. Result of pressure and temperature reaction of curing concrete with CO_2 for one hour

the age of concrete but it may vary for different CO_2 concentration, duration and pressure combinations.

3.3 Concrete Compressive Strength Associated with Cylinder Size

The standard size specimen used for compressive strength acceptance testing is a 15 × 30 cm cylinder. ASTM, AASHTO or CSA standards allow the use of 10 × 20 cm cylinders or 5 × 10 cm cylinders. However, these cylinders are not often used because of the uncertainty of how their strength compares to the strength of 15 × 30 cm cylinders made from the same batch of concrete. This section studies that have been done to correlate the strengths between a standard cylinder size (15 × 30 cm) and one that is smaller. The strength results and strength conversion factors (ks value) obtained from 5 × 10 cm, 10 × 20 cm, and 15 × 30 cm cylinders by water curing are displayed in Table 3. Each strength value is the mean of three cylinder strengths. It was found that k_{s5} and k_{s10} obtained from 3-day compressive strength were almost equal to 1. There was no difference in strength between 10 × 20 cm, and 15 × 30 cm cylinders and its k_{s10} ranges between 0.97 and 1.03. It should be noted that there was difference in strength between 5 × 10 cm, and

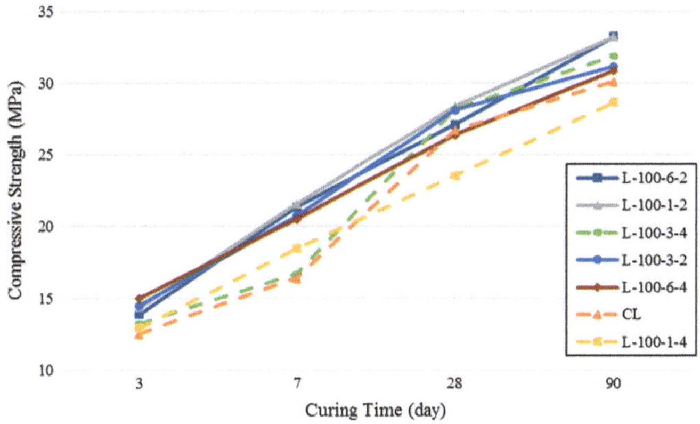

(a) 100% CO_2 concentration

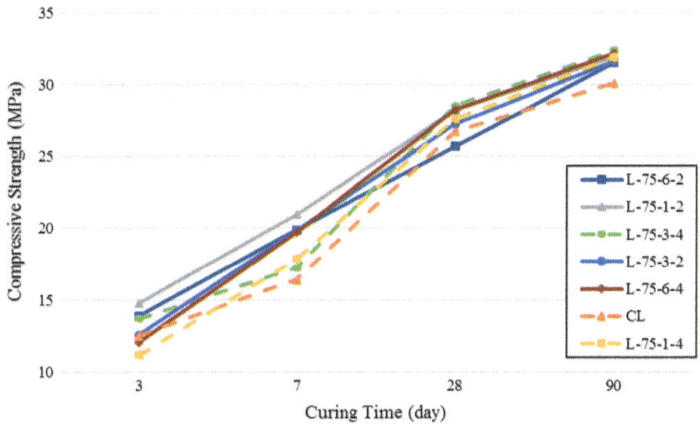

(b) 75% CO_2 concentration

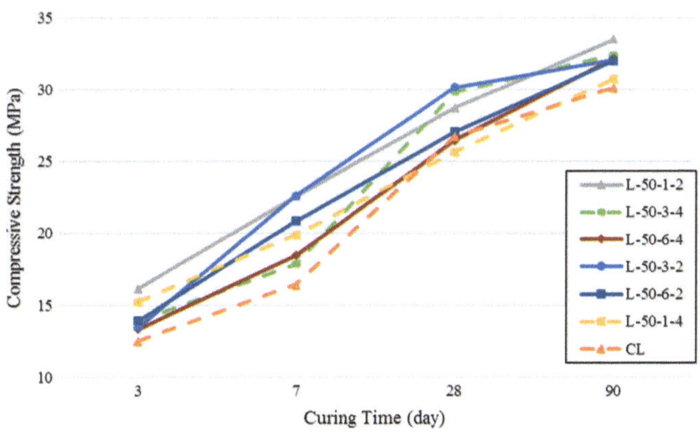

(c) 50% CO_2 concentration

Fig. 3. CO_2 concentration and curing variables on compressive strength for 15 × 30 cm cylinder

15 × 30 cm cylinders and its k_{s5} obtained after 7-day compressive strength was greater than 1.13.

Table 3. Strength and k_s value obtained from three size cylinders by water curing (unit: MPa)

Cylinder size	Φ5 × 10 cm	Φ10 × 20 cm	Φ15 × 30 cm	k_{s5} (f_{c5}/f_{c15}), k_{s10} (f_{c10}/f_{c15})
3 day	12.34	12.49	12.53	0.98, 0.99
7 day	19.33	15.90	16.42	1.18, 0.97
28 day	31.59	27.65	26.73	1.18, 1.03
90 day	34.06	30.02	30.12	1.13, 1.00

3.4 Concrete Compressive Strength Associated with CO_2 Curing and Cylinder Size

For normal strength concrete, the two different cylinder (10 × 20 cm, and 15 × 30 cm) sizes do not result in differences in compressive strength tests performed at 3, 7, 28, and 90 days from the above section. However, it was observed that the size effect between 5 × 10 cm and 15 × 30 cm cylinders became substantially in high strength concrete [11]. The strength conversion factors (k_{s5} value) obtained from 5 × 10 cm, and 15 × 30 cm cylinders by CO_2 curing are displayed in Fig. 4. It was found that there was difference in strength conversion factors between 5 × 10 cm, and 15 × 30 cm cylinders. And the k_{s5} value obtained at 3-day compressive strength ranges between 0.99 and 1.30. For example, a cylinder with the identity L-50-1-2 made from the general strength of concrete with 50% CO_2 concentration curing 1 hour under 2-bar pressure has a high k_{s5} value obtained at 3-day compressive strength was about 1.30 as compared to the baseline water-cured specimens (k_{s5} value 0.98). This is partly due to the fact that high early compressive strength of the CO_2-cured cylinders was obtained while its early strength may have achieved the 28-day strength of normally water-cured ones. And the k_{s5} value obtained at 90-day compressive strength ranges between 0.95 and 1.25 as close to the baseline water-cured specimens (k_{s5} value 1.13). The order of strength data in the age of concrete may vary for different CO_2 concentration, duration and pressure combinations. Therefore, special care is still needed to compare the properties of concrete from the two different cylinder sizes by CO_2 curing. It can be seen that the CO_2-cured samples have slightly higher k_{s5} value and variation as compared with the strength of normally water-cured ones (CL) at 3-day curing duration.

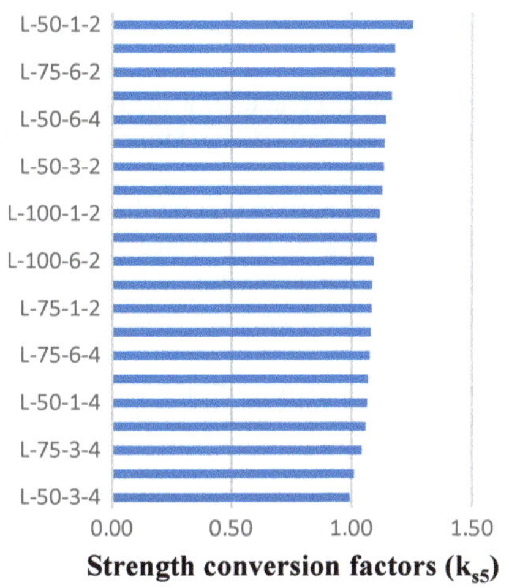

(a) 3-day compressive strength

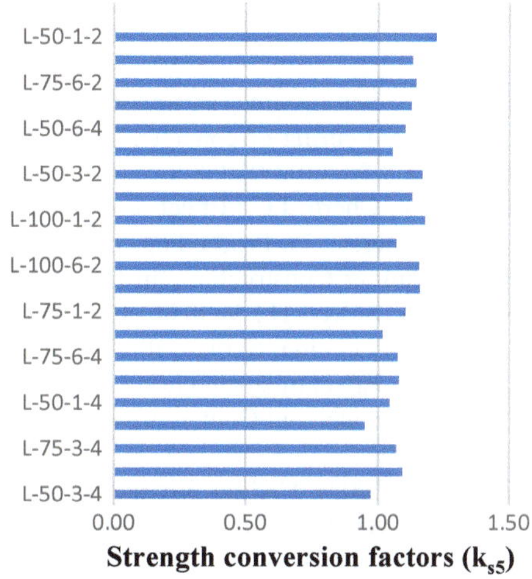

(b) 90-day compressive strength

Fig. 4. Result of strength conversion factors (k_{s5} value) of concrete CO_2 curing

4 Summary

The main findings from this study are summarized as follows:

(1) The performance of the CO_2-cured cylinders was found through the measurement of pressure drop, the temperature expected to go down and rise, and CO_2 was being absorbed by concrete.

(2) Most of the CO_2-cured cylinders have slightly higher compressive strength as compared with the baseline water-cured ones at curing times ranging from 3 to 90 days. It is good to hear that the strength of concrete increases with an increase in the age of CO_2-cured concrete but it may vary for different CO_2 concentration, duration and pressure combinations.

(3) For water cured concrete, there were no differences in 3, 7, 28, and 90-day strengths between 10 × 20 cm, and 15 × 30 cm cylinders and their strength conversion factor (k_{s10}) close to 1. There are differences in strength between 5 × 10 cm, and 15 × 30 cm cylinders and its k_{s5} obtained after 7-day compressive strength was greater than 1.13.

(4) The CO_2-cured cylinder has a high k_{s5} value and variation obtained at 3-day compressive strength as compared to the baseline water-cured ones. This is partly due to the fact that high early compressive strength of the CO_2-cured cylinders was obtained while its early strength may have achieved its *28 day* value.

References

1. Erdem, T.: Setting time: an important criterion to determine the length of the delay period before steam curing of concrete. Cem. Concr. Res. **33**, 741–750 (2003). https://doi.org/10.1016/s0008-8846(02)01058-x
2. Lee, M.G., Chern, J.C.: Effect of steam-curing cycle and proportion mix on the strength of precast concrete. In: Proceedings of the Seventh East Asia-Pacific Conference on Structural Engineering & Construction, pp.1390–1395. JSCE, Kochi, Japan (1999)
3. Kosmatka, Steven H., William, C.P.: Design and Control of Concrete Mixtures. Portland Cement Association, Skokie, Illinois (1988)
4. Lee, M.G.: Effect of steam-curing on the strength of precast concrete. In: Proceedings of the Seventh International Conference on Composites Engineering, pp. 513–514. ICCE, Denver, USA (2000)
5. Gang Ye (2003). Carbon dioxide uptake by concrete through early age curing. Master's thesis, University of McGill University, Canada
6. Zhan, B., Poon, C.S., Liu, Q., Kou, S., Shi, C.: Experimental study on CO_2 curing for enhancement of recycled aggregate properties. Constr. Build. Mater. **67**, 3–7 (2014)
7. Kou, S., Zhan, B., Poon, C.S.: Use of a CO_2 curing step to improve the properties of concrete prepared with recycled aggregates. Cem Concr Compos **45**, 22–28 (2014)
8. Zhan, B., Xuan, D.X., Poon, C.S., Shi, C.: Effect of curing parameters on CO_2 curing of concrete blocks containing recycled aggregates. Cem Concr Compos **71**, 122–130 (2016)
9. Zhan, B., Poon, C.S., Shi, C.: CO_2 curing for improving the properties of concrete blocks containing recycled aggregates. Cem Concr Compos **42**, 1–8 (2013)
10. Zhan, B., Poon, C.S., Shi, C.: Materials characteristics affecting CO_2 curing of concrete blocks containing recycled aggregates. Cem Concr Compos **67**, 50–59 (2016)
11. Day, R.L.: Strength measurement of concrete using different cylinder sizes: a statistical analysis. Cem. Concr. Aggregates **16**(1), 21–30 (1994)
12. Malhotra, V.M.: Are 4 by 8-in concrete cylinders as good as 6 by 12-in cylinders for quality control of concrete? ACI J. **73**(1), 33–36 (1976)
13. ASTM C39/39 M-16: Standard Test Method for Compressive Strength of Cylindrical Concrete Specimens, ASTM International, West Conshohocken, PA. www.astm.org (2016)
14. Lee, M.G., Wang, W.C., Huang, SU, Y.M., Jiang, Q.: Effect of carbon dioxide curing on strength development of cement mortar. In: Key Engineering Materials, vol. 748, pp. 323–327 (2017)

Performance Study on Asphalt Pavement Preservation Technologies in Gansu Province

Dingbang Wei[1], Xiaomin Li[2], and DingXin Cheng[3(✉)]

[1] Lanzhou Jiaotong University, Lanzhou, Gansu, China
286237855@qq.com

[2] Gansu Institute of Highway Maintenance Technology, Lanzhou, Gansu, China
hitlxm@126.com

[3] California Pavement Preservation Center, California State University, Chico, USA
dxcheng@csuchico.edu

Abstract. Pavement preservation majorly includes preventive maintenance, routine maintenance, and minor rehabilitation. Pavement preservation has many benefits when utilized correctly including life extension of the existing pavement, low treatment costs, reduced user costs, improved safety and overall network health, etc. A strong preservation program is essential to asset management because it is more cost effective to address minor distresses early, before the defects become major costly problems. Pavement preservation consists of many different kinds of treatments; however, each location has its suitable treatments, particularly in those complex climate environment. Gansu province has various geological and climate conditions, which create challenges to pavement engineering. To find the suitable pavement preservation treatments, a series of pilot testing sections using various pavement preservation technologies have been set up for a long-term pavement performance (LTPP) study. This comprehensive study involves four climate regions (Hexi desert region, Longnan rainy regions, Gannan alpine region and Longdong loess region), six types of general observation sections (Asphalt Concrete, SUPERPAVE, Asphalt Treat Base and so on), and five types of special preservation sections (SPS) (micro-surfacing, cape seal, ultrathin overlay, rubberized asphalt, and Stone mastic asphalt mixture). These test sections are used to help treatment design, strategy selection, and long-term performance modeling of pavement preservation treatments. For the test sections, pavement performance measures, such as pavement distresses, pavement condition index (PCI), Rutting Depth Index (RDI), are observed quarterly each year. To further support the analysis, weather stations have been established to collect temperatures and humidity information for pavements. The collected data are used to determine the influence of different maintenance technologies on pavement performance. The current study results have shown that pavements are more susceptible to crack and rut in the Hexi large temperature variation areas, and the cape seal technology and rubberized asphalt overlay are more suitable for retarding reflective cracking.

Keywords: Pavement preservation · Maintenance technology
Long term pavement performance · Pavement condition

M. Barman et al. (eds.), *Transportation and Geotechniques: Materials, Sustainability and Climate*, Sustainable Civil Infrastructures,
https://doi.org/10.1007/978-3-319-95768-5_5

1 Introduction

To evaluate the pavement distress condition comprehensively, the collected data from pavement performance need to be as detailed as possible. Also, it is better to collect the distress data from the beginning of pavement service so that the distress deterioration curve can be analyzed. Besides, pavement preservation technologies were used widely nowadays to extend pavement life, and the effectiveness was evaluated by many researchers.

Currently, the most complete pavement distress database is from the Long Term Pavement Program (LTPP) of the United States, a part of Strategic Highway Research Program (SHRP) initiated in the 1980s. Since the 1990s, LTPP data has been extensively used by researchers to evaluate pavement performance or treatment effectiveness (Elkins et al. 2003; Hall et al. 2002; Rauhut et al. 2000). A variety of performance and distress indicators were investigated, including fatigue cracking, longitudinal cracking in wheel path and not in wheel path, transverse cracking, rutting and roughness (Chen et al. 2003; Dong and Huang 2012a, b, 2014, 2015; Dong et al. 2013; Haider and Chatti 2009; Hall et al. 2002, 2003; Karlaftis and Badr 2015; Wang and Wang 2013; Wang et al. 2005).

Meanwhile, in Australia, Long Term Pavement Performance Maintenance (LTPPM) sites were included in the program to: (i) assess the impact of surface maintenance treatment on changes to pavement conditions (works effects, WE); and, (ii) assess the impact of maintenance on road deterioration (RD) (Martin and Choummanivong 2016).

However, with so many types of distress data collected in LTPP, several questions raised, including how to utilize those data to recognize the rule of pavement performance changes. It is noted that LTPP SPS-5 data have been used for several studies. National Cooperative Highway Research Program (NCHRP) project 20-50 reported that the performance of various pavement maintenance and rehabilitation treatment techniques is compared by using LTPP data, including the SPS-5 data. Wang introduced the field performance of thin and thick overlay using Reclaimed Asphalt Pavement (RAP). He found that RAP-containing mixtures outperform virgin mixtures in rutting and roughness, without inducing additional cracking-related distress (Wang 2016). Hao Wang compared the effectiveness of preservation treatments on pavement surface friction and investigated the long-term variation of friction using the data collected in the SPS-3 of the LTPP program. He found that the application rate of slurry seal and chip seal affected the friction variation and temperature showed negative correlation with friction (Wang et al. 2013). Furthermore, the effective time for applications of slurry seal on asphalt pavement was studied by Elie that the optimum times and highest relative benefit of slurry seals for both overlaid and newly constructed pavement were when the first slurry was applied at 3 years after construction, followed by a second slurry seal at year 7 or 9 (Hajj et al. 2013).

As for pavement crack, data from SPS-5 of LTPP was also used for prediction model development. Aristides G indicated that the proposed approach results in accurately predicting the probability of crack initiation following treatment;

Furthermore it provided information on the relationship between external factors and cracking probability that can help pavement managers in developing appropriate rehabilitation strategies (Karlaftis and Badr 2015).

1.1 LTPP Set up in Gansu Province, China

Located in the northwest of China, Gansu province is the strategic channel connecting Mid-Asia and Europe. With the development of highway construction, many efficient preservation technologies in Gansu province have been widely used and developed, including micro-surfacing, chip seal, cape seal, ultrathin overlay, regeneration technology, and so on. However, these technologies need further research.

In addition, climate condition in Gansu is complex, which is divided into four regionalizations (Hexi desert region, Longnan rainy regions, Gannan alpine region and Longdong loess region, as shown in Fig. 1), and different treatments adopted according to different climate regions, especially those regions having large temperature range. Many pavement preservation treatments were used widely in cold regions, chip seal, micro-surfacing and thin overlay have the longest service life (Zubeck et al. 2012), but few results can we found in large temperature-difference regions.

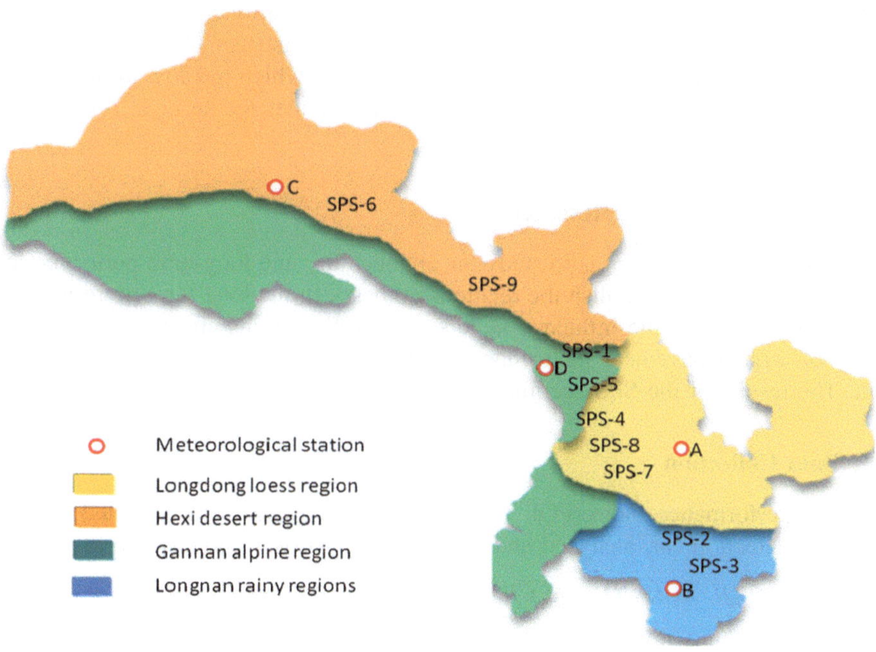

Fig. 1. Four climate regions in Gansu province

1.2 Pavement Preservation Technology

The following are the types of pavement preservation technologies selected for this study:

Micro-surfacing: A cold mixture includes polymer modified asphalt emulsion, aggregate, mineral filler, additives and water. Micro-surfacing can be applied to recovery skid resistance or fill rut. Type 3 Micro-surfacing mixture (maximum aggregate size = 13.5 mm) of 10.5% asphalt emulsion content was applied in this study.

Cape seal: A composite seal make up with chip seal and micro-surfacing, which can treat the reflection crack. Polymer modified asphalt (1.5 kg/m^2 spraying volume) and crushed aggregate (9.5 mm size) were applied in this study.

Ultrathin overlay: A thin overlay about 2.5 cm applied on pavement surface aimed to treat surface damage. High viscosity asphalt and quick setting asphalt emulsion were applied in this study.

ARHM: Asphalt Rubber Hot Mixture is a type of hot mixture using asphalt. Usually, crumb rubber content is about 20% by weight of asphalt. ARHM is an effective technology to treat reflection crack.

SMA: Stone Mastic Asphalt is a type of hot mixture which has high content of asphalt, filler, and coarse aggregate. SMA has a property of high strength and fatigue resistance.

SUPERPAVE: A type of hot mixture designed by gyratory compaction, which has good property of rut resistance and moisture damage resistance.

ATB: Asphalt treated base is a type of hot mixture which has low asphalt content and large aggregate size (up to 30 mm). As a layer under the surface, ATB offers a powerful support to the pavement.

1.3 The Purpose of the Research

The purpose of this research is to study the applicability and long-term performance of asphalt pavement, and establish the relationship between the long-term performance of asphalt pavement and the influence factors. The SPS sections aim to study the technology of asphalt pavement maintenance and the performance decay rule after treatment. The results of the SPS section selection are shown in Table 1.

1.4 Data Collection

Pavement performance index PQI, PCI, RDI and CR data were collected by Road Intelligent detection car, analysed and calculated by software as show in Fig. 2.

Definition of PQI, PCI, RDI and CR

According to Chinese transportation standard <Highway Performance Assessment Standards> (JTJ H20-2007), pavement performance index was defined as:

- PQI-Pavement Quality or Performance Index: A general index including pavement strength, damage and evenness.

Table 1. SPS sections

Number	Maintenance technology	Year build	Year maintenance	Layer	History
SPS-1	MS-3	2013	2013	Surface	Skid resistance decay
SPS-2	MS-3	2015	2015	Surface	Skid resistance decay
SPS-3	Novachip	2014	2014	Surface	Skid resistance decay
SPS-4	Novachip	2015	2015	Surface	Moisture damage
SPS-5	Capseal	2015	2015	Surface	Reflection crack
SPS-6	5 cm ARHM-13	2013	2013	Surface	Reflection crack
SPS-7	4 cm Super-13	2014	2014	Surface	Rutting
SPS-8	5 cm SMA-13+10 cm ATB-25	2014	2014	Surface	Moisture damage
SPS-9	4 cm Super-13+8 cm ATB-25	2014	2014	Surface	Rutting

Fig. 2. Road intelligent detection car

- PCI-Pavement Surface Condition Index: A general index including crack, pot holes and rut and so on.
- RDI-Rutting Depth Index.
- CR-Cracking Rate: The crack area (The crack length by 0.2 m) in 100 m pavement.

2 Results and Analysis

2.1 The Temperature Characteristics of Different SPS Sections in Gansu Province

In order to observe the temperature characteristics of four climate regions in Gansu province, four meteorological stations are distributed in 9 SPS sections. Nearly half a year meteorological data (including winter and summer) were collected. As shown in Figs. 3, 4, 5 and 6, from the mean value of the view, Lanzhou and Qingshui belong to high-temperature region, with extreme value reaching 40.41 °C. In addition, Wushaoling and Jiayuguan belong to low-temperature area with the low-temperature

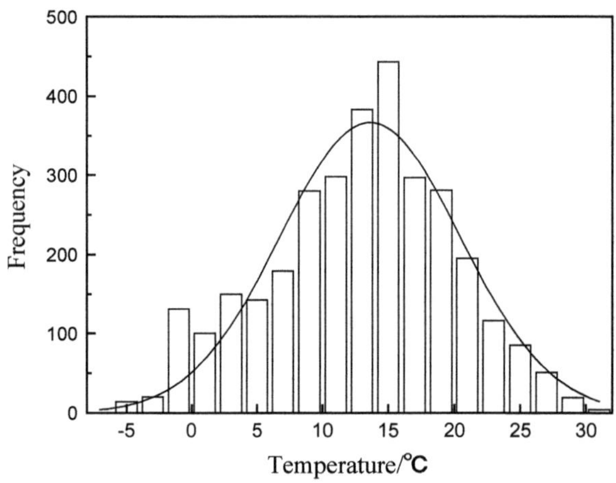

Fig. 3. Normal distribution curve (station D)

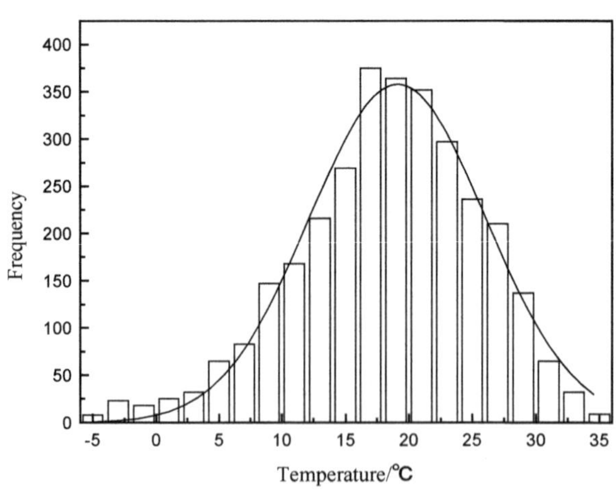

Fig. 4. Normal distribution curve (station A)

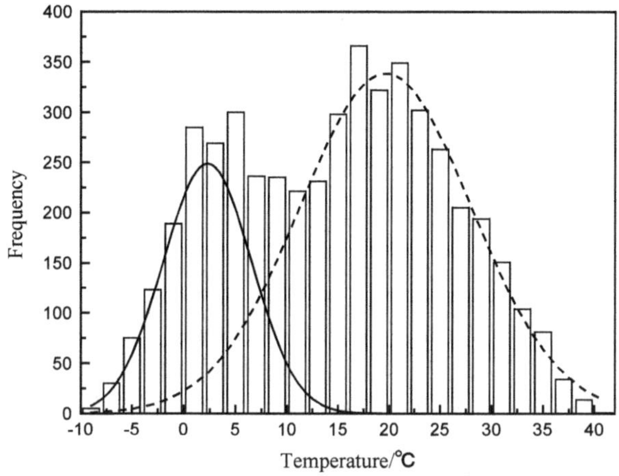

Fig. 5. Normal distribution curve (station B)

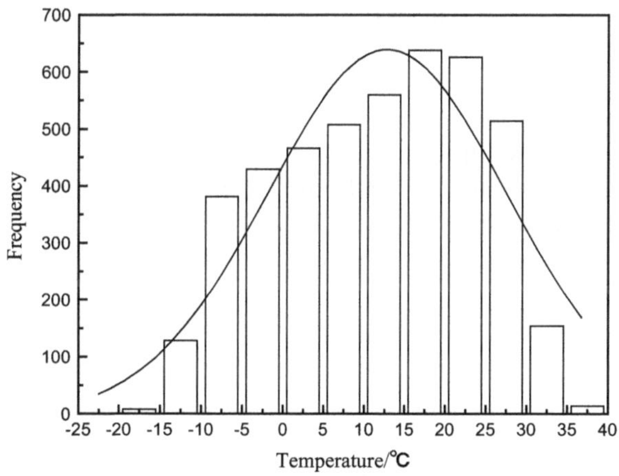

Fig. 6. Normal distribution curve (station C)

extreme value reaching −20.08 °C. Among them, Jiayuguan is a region with large temperature difference, and the temperature range is 55.88 °C. Because of the large temperature difference, maintenance materials applied in these regions need good properties in both high and low temperatures (Table 2).

2.2 The Variation of Pavement Treatments Performance in Different Sections

The long-term observation index data of PQI, PCI, RDI and CR of the 9 SPS sections were collected. Then, through the comparative analysis of the annual decay regulation

Table 2. Meteorological station data analyze

Meteorological station	Max	Min	Ave	Std	Temperature range
Lanzhou	35.48	−5.17	18.3	7.14	40.65
Tianshui	40.41	−8.48	14.78	10.59	48.89
Jiayuguan	35.8	−20.08	11.53	12.07	55.88
Wushaoling	30.8	−5.95	12.84	6.98	36.75

of the pavement condition index, the effect of different maintenance technology is analyzed, the results are shown in Figs. 7, 8, 9, 10 and 11.

As shown in Figs. 7, 8, 9, 10 and 11, the pavement condition of 9 SPS sections using different maintenance techniques show a decline in PQI, PCI, RDI, CR year by year.

As far as preventive maintenance is concerned, the pavement condition index with micro-surfacing technology (SPS-1, SPS-2) decay rapidly. The rate of decay of PCI reached 27.8% in 4 years, and the CR, increased to 13.3% in 4 years. The pavement condition index with ultrathin overlay technology has a lower attenuation, which maintained at 5%. Particularly, the CR increased less than 2 times. And the cape seal technology, with little PQI attenuation and lower CR growth, has better pavement performance.

In terms of the corrective maintenance technology, the ARHM-13 overlay with rubberized asphalt and SUPER-13 overlay (SPS-6, SPS-7) can maintain a better pavement condition, especially the CR of ARHM-13 only increased 0.91 times. However, the attenuation of both the PCI and RDI of pavement, overlaid by 5 cm SMA-13+10 cm ATB-25 and 4 cm Super-13+8 cm ATB-25 (SPS-8, SPS-9) were greater.

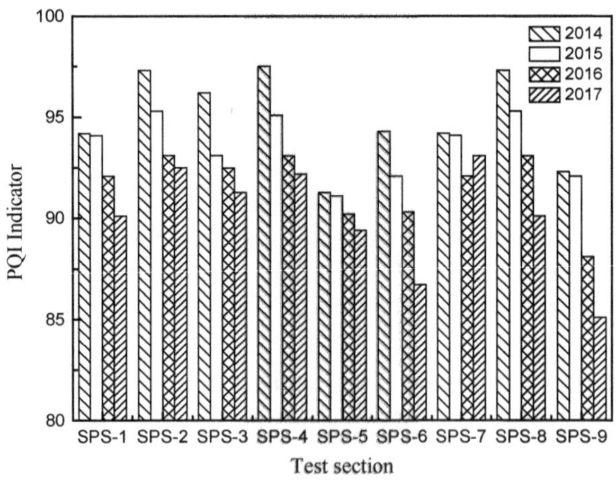

Fig. 7. PQI decay with year of different sections

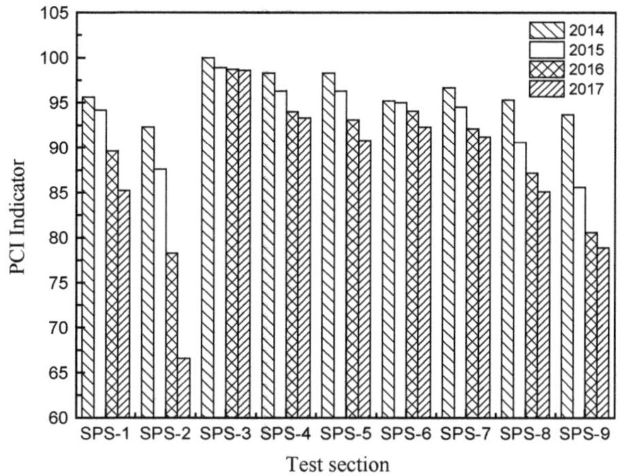

Fig. 8. PCI decay with year of different sections

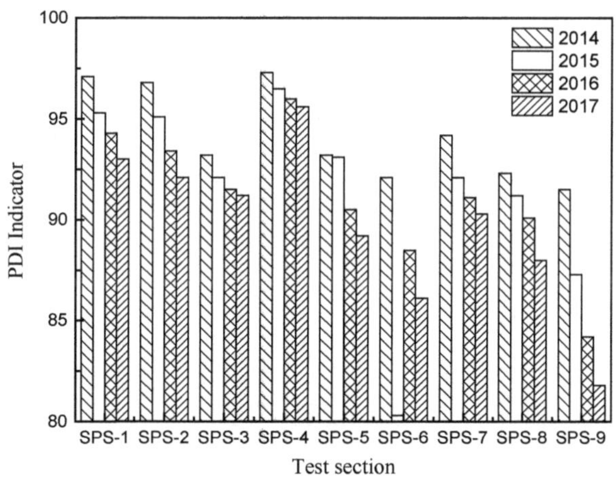

Fig. 9. RDI decay with year of different sections

Through the 5 figures and results analysis above, we can draw a conclusion that ultrathin overlay and chip seal give a better performance compared with micro-surfacing, especially in the maintenance of surface function.

2.3 Characteristics of Cracks/Rutting in Large Temperature Difference Areas

There is a close relationship between temperature or temperature difference and surface crack rate/rutting during pavement service. As a typical viscoelastic materials, asphalt is sensitive to temperature and subject to crack at low temperatures and rut at high temperatures.

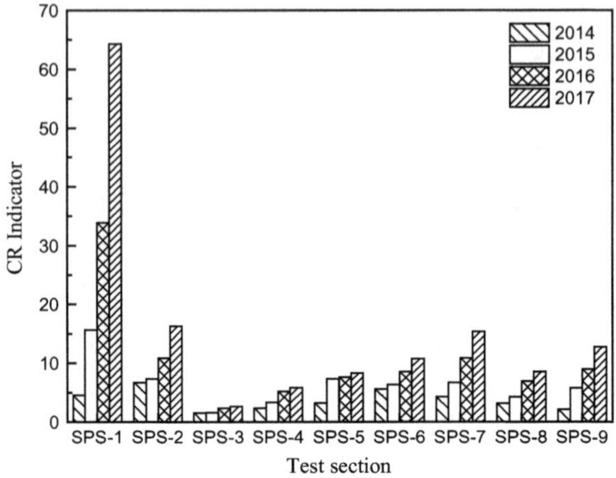

Fig. 10. CR decay with year of different sections

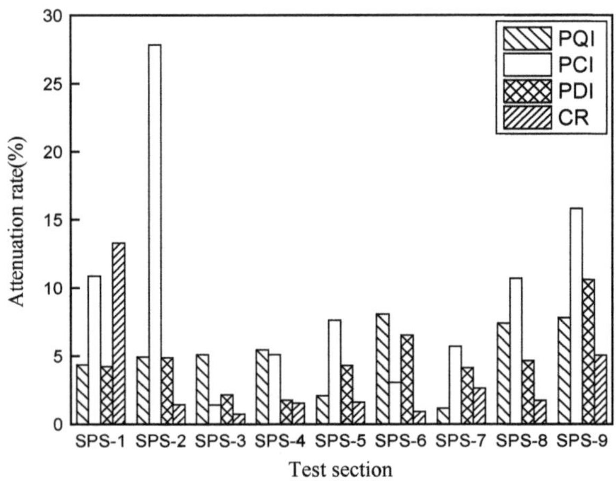

Fig. 11. Decay of 4 index of different sections

It can be seen from the Fig. 12 that the values of RDI and CR have some growth trends with the increase of the temperature range (temperature difference), indicating that the asphalt pavements are more prone to cracking and rutting in the large temperature difference area (such as the Hexi Gobi region). All phenomena conform to the characteristics of pavement damage due to large temperature difference in Gansu province.

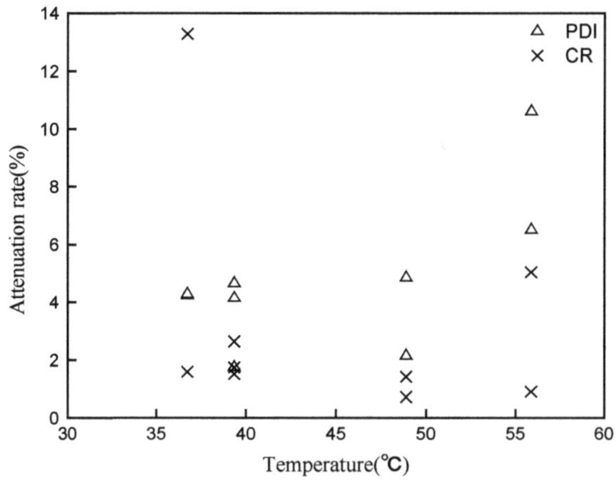

Fig. 12. RDI and CR decay with year of different sections

2.4 The Results of Anti-cracking Based on Different Maintenance Technologies

Reflective cracking is a common distress of semi-rigid base layer pavement, caused by cracking of semi-rigid base layer. Overlay and other technologies can effectively reduce the reflection of cracks.

It can be seen from the Fig. 13 that the rubberized asphalt overlay, cape seal and Novachip overlay have a significant effect on anti-cracking. And it turned out to be working well. All crack growth of different maintenance technologies are below 2 during the 4 years, in which crack growth of rubberized asphalt overlay is only 0.91 times.

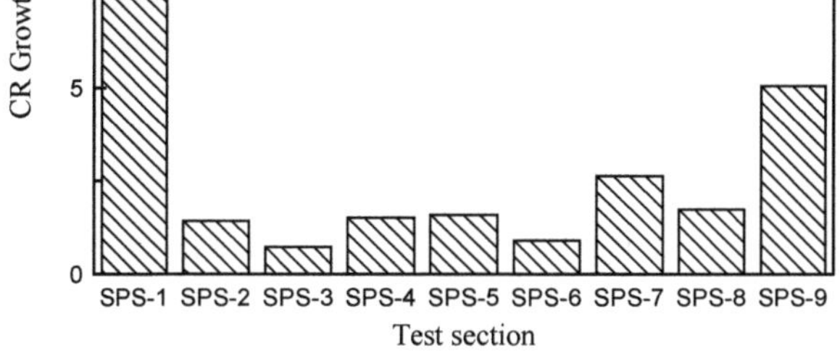

Fig. 13. CR decay of different sections

3 Conclusions

The following are the preliminary conclusions from the study:

- Gansu province has set up a LTPP program involving four climate regions, six types of general observation sections, and five types of special observation sections. The LTPP program will support pavement design, strategy selection, and long-term performance modeling.
- Different treatments performed differently in Gansu province. Micro-surfacing decay rapidly, while ultrathin overlay and cape seal performed well. The rubberized asphalt and SUPER-13 overlay maintained a good pavement condition for corrective maintenance.
- The study approves that large temperature differences can cause high pavement distresses such as rutting and cracking. The rubberized asphalt overlay, cape seal and Novachip overlay have performed well to resist reflective cracking from the semi-rigid base layer.
- The research results of LTPP in Gansu province provide effective data support for the selection of highway maintenance technology in high altitude and large temperature difference areas.

Acknowledgments. This study was a part of the project funded by the Administrator of Gansu Transportation.

References

Gautam, D., et al.: Common structural and construction deficiencies of Nepalese. Innov. Infrastruct. Solut. (2016). https://doi.org/10.1007/s41062-016-0001-3

Elkins, G.E., Schmalzer, P., Thompson, T., Simpson, A.: Long-term Pavement Performance Information Management System Pavement Performance Database User Reference Guide (2003)

Hall, K.T., Simpson, A.L., Corea, C.E.: LTPP Data Analysis Effectiveness of Maintenance and Rehabilitation Options. National Cooperative Highway Research Program (2002)

Rauhut, B., Von Quintus, H., Eltahan, A.: Performance of Rehabilitated Asphalt Concrete Pavement in LTPP Experiments (Data collected Through February 1997). Publication No. FHWA-RD-00-029, June (2000)

Chen, D., Lin, D., Luo, H.: Effectiveness of preventative maintenance treatments using fourteen SPS-3 sites in Texas. J. Perform. Constr. Facil. **17**(3), 136–143 (2003)

Dong, Q., Huang, B.: Evaluation of effectiveness and cost-effectiveness of asphalt pavement rehabilitation utilizing LTPP data. J. Transp. Eng. **138**(6), 681–689 (2012a)

Dong, Q., Huang, B.: Evaluation of influence factors on crack initiation of LTPP resurfaced-asphalt pavements using parametric survival analysis. J. Perform. Constr. Facil. **28**(2), 412–421 (2012b)

Dong, Q., Huang, B.: Analyzing Failure Probability of LTPP Pavement preventive maintenance treatments though survival analysis. Transportation Research Board 94th Annual Meeting (2015)

Dong, Q., Jiang, X., Huang, B., Richards, S.H.: Analyzing influence factors of transverse cracking on LTPP resurfaced asphalt pavements through NB and ZNB models. J. Transp. Eng. **139**(9), 889–895 (2013)

Haider, S.W., Chatti, K.: Effect of design and site factors on fatigue cracking of new flexible pavements in the LTPP SPS-1 experiment. Int. J. Pavement Eng. **10**(2), 133–147 (2009)

Hall, K., Correa, C., Simpson, A.: Performance of flexible pavement rehabilitation treatments in the long-term pavement performance experiment. Transp. Res. Rec. J. Transp. Res. Board **1823**, 93–101 (2003)

Karlaftis, A.G., Badr, A.: Predicting asphalt pavement crack initiation following rehabilitation treatments. Transp. Res. Part C Emerg. Technol. **55**, 510–517 (2015)

Wang, H., Wang, Z.: Evaluation of pavement surface friction subject to various pavement preservation treatments. Constr. Build. Mater. **48**, 194–202 (2013)

Wang, Y., Mahboub, K.C., Hancher, D.E.: Survival analysis of fatigue cracking for flexible pavement based on long-term pavement performance data. J. Transp. Eng. **131**(8), 608–616 (2005)

Martin, T., Choummanivong, L.: The benefits of Long-term Pavement Performance(LTPP) research to funders. Transp. Res. Procedia **14**, 2477–2486 (2016)

Wang, Y.: The effect of using reclaimed asphalt pavement (RAP) on the long-term performance of asphalt concrete overlays. Constr. Build. Mater. **120**(2016), 335–348 (2016)

Hajj, E.Y., Luis, L.G., Sebaaly, P.E., Cortez, E.: Effective timing for two sequential applications of slurry seal on asphalt pavement. J. Transp. Eng. **139**(5), 476–484 (2013)

Zubeck, H., Mullin, A., Liu, J.: Pavement preservation practices in cold regions. Cold Regions Engineering 2012: Sustainable Infrastructure Development in a Changing Cold Environment (2012)

A New Procedure for Construction Metro Station in Non-cemented Soil: A Numerical Investigation

Mahsa Firouzi and Hamed Bayesteh(✉)

Department of Civil Engineering, University of Qom, Qom, Iran
h.bayesteh@qom.ac.ir

Abstract. According to the development of societies and increase of population, the use of underground spaces has been developed in the past decades. The high cost of construction of the urban space such as metro stations, leads to increase the demand for tunnels in big urban. Thus, an economical method should select to build these structures. One of the popular and flexible underground construction methods is New Austrian Tunneling Method (NATM). When NATM is implemented in big cross section, face is divided in to temporary section. This method includes Central Diaphragm (CD) method, Side wall Drift (SD) method, Pile-Beam-Arch method and etc. In the non-cemented soil, the settlements due to the excavation sequences maybe leads to damage to nearby structures. Therefore, it is necessary to investigate the effective factor of controlling tunnel-induced settlements and change the construction stages during excavation. A lot of numerical studies have been done on NATM in urban areas that often deal with the simulation of tunnels and small sections. In this paper, a number of Finite Difference Method analyses were conducted to evaluate the effects of different patterns for advancing the tunnel face on the settlement in NATM station construction in the non-cementation soil. Constructing a station is considered as a big section. Induced displacements are empirically controlled by adjusting the speed of excavation, installing support and partial-face excavation.

1 Introduction

With increasing urban population and increasing transportation problems, the construction of underground spaces and subway is developing. These spaces are often built close to the ground to make passengers easier to access and reduce investment and operational costs. These structures are often built in non-cemented ground and with a low overburden, which will cause sensitivities and problems. The control of settlement in urban areas is very important. The need for settlement control has been widely known and new construction methods have been developed (Farias et al. 2004). A key factor in the successful completion of a project in the tunneling in metropolitan cities is the proper design and construction method of a tunnel, which will greatly affect the costs and completion time of the project.

Generally, tunnel construction can be divided into two mechanized and traditional categories. Several factors are involved in the selection of each of these methods, such

M. Barman et al. (eds.), *Transportation and Geotechniques: Materials, Sustainability and Climate*, Sustainable Civil Infrastructures,
https://doi.org/10.1007/978-3-319-95768-5_6

as project finance, technical experience of the design and implementation team, environmental issues, etc. (Sharifzadeh et al. 2013). In these days the use of TBM is developing, but because of the high cost of installation, it is profitable for tunnels with considerable length. Consequently, traditional methods use in difficult conditions, short height spaces and variable sections (Kolymbas 2003).

Because stations are short height spaces, they are constructed by traditional methods. In these category, there are two methods for constructing the underground station: cut & cover and underground methods. If there is enough space on the ground and the station is shallow cut & cover is selected otherwise underground method is used. NATM is a flexible method that is compatible with complex and various conditions (Sharifzadeh et al. 2013) so it is suitable for constructing stations. In this method, generally face is excavated in multi-stage because the section of station is large.

The three principles of NATM are: 1. Soil or rock mass should be considered an active part of the tunnel that can sustain part of load. Soil conditions may improve by grouting, anchor and so on. To reduce the stress, the shape of the tunnel should be optimized. Usually, to reduce the settlement. 2. Lining should be installed optimally to complete support capacity and reduce deformation. Lining should work like a thin wall and be flexible and strong enough to absorb the shear and moments. Shotcrete is suitable for this purpose. 3. The third principle relates to the instrument. Design is based on experience, but the excavation method may change during construction based on measured data (Farias et al. 2004).

A lot of numerical studies have been done on NATM in urban areas that often deal with the simulation of tunnels and small sections. In this study, a numerical method has been used to simulate a station with large section using underground excavation method, and the results have been presented.

2 Geological and Geotechnical Specification

Here a case study is simulated using underground method. The station section has a width of 17.4 m and a length of 18.4 m. According to studies, Soil in this area is predominantly SC, SM, GP-GM, GC, GM and GW. In this site, according to the standard penetration and results of laboratory tests and statistical surveys, regardless of the first 4-m soil, the soil is relatively homogeneous. The characteristics of the soil layers are given in Table 1.

Table 1. Characteristics of the soil layers

Depth (m)	γ (kPa)	Modulus secant (MPa)	Modulus unloading (MPa)	Poisson ratio	Cohesion (kPa)	Internal friction angle (°)	K0
0–4	17	20	60	0.2	10	29	0.52
4–18	19.5	150	750	0.2	40	38	0.38
18–23	20	200	1000	0.2	10	40	0.36
>23	20.7	250	1250	0.2	30	39	0.37

3 Construction Process

The station is constructed by NATM and because of large cross section, it is excavated partially. Different methods of excavating have been developed in this way, like Side Drift, Ring-Cut and Central Diaphragm methods. In order to determine the sequences of excavation, according to the soil conditions and stress ratios, Yu and Chern (2007) proposed the diagram in Fig. 2 based on the span size and the ratio of axial compressive strength to vertical stress enters the tunnel. Based on this diagram, the Side Drift method is selected for excavating (Fig. 1).

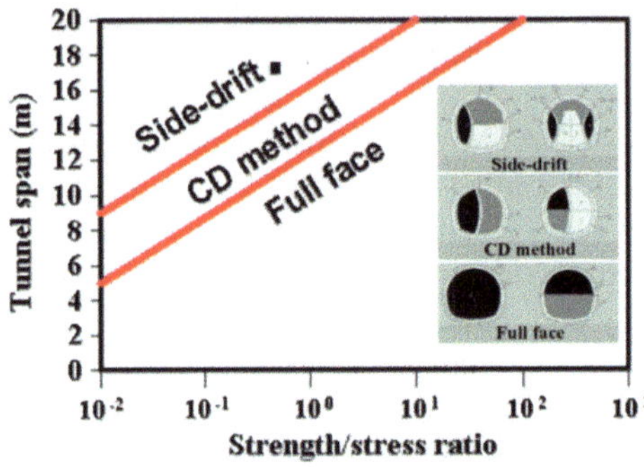

Fig. 1. Diagram for selecting a suitable excavation method [9]

The construction process includes excavation and installing a temporary support. First, the drifts are excavated (Fig. 2a). After excavating the upper stage, the temporary lining is implemented. The temporary lining is included shotcrete, two set of mesh and, lattice girder in 1 m distances. Shotcrete of walls has a thickness of 40 cm and the temporary invert with a thickness of 15 cm (Fig. 2b). In this stage, 6 length nail and micropiles are installed in 1 m distances. Micropiles are 8 and 10 m length and include one rebar with 25 mm diameter and casing with 76 mm diameter and 5 mm thickness. After it, top heading is excavated and supported. The excavation continues as it is shown in the Fig. 4. At the end, bench and invert are excavated and supporting ring is closed (Fig. 2e to h). The nails used at one side because there is a structure at the right side of the station.

4 Simulation of Finite Element

Based on the nature of the problem and the distribution of stresses and displacements numerical analysis includes two-dimensional and three-dimensional analyzes (Sharifzadeh et al. 2013). In this study, a cross-section was selected in the middle of the station and in a place that there aren't any intersections. It is assumed that there is the plane strain condition and it is simulated by FLAC 2D software. It is explicit finite difference program and plastic hardening model, used here, is a new model provided in it.

In many cases, specific assumptions consider the effect of face. In this paper by using convergence-confinement method (stress relief method) the effect of the third

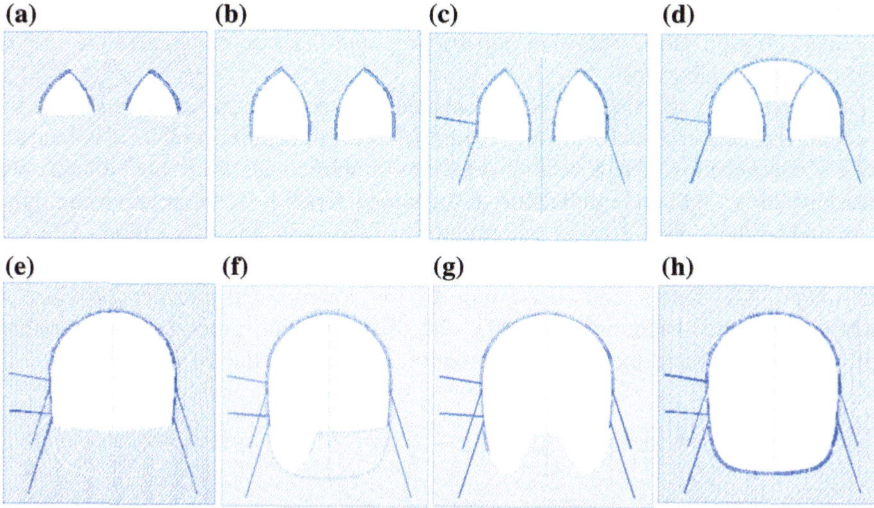

Fig. 2. Construction process (a–h)

dimension is considered. In this theory, the stress entered into the tunnel is gradually reduced, and by plotting the reduction factor versus distance from the face, the appropriate reduction factor is obtained (Karakus 2007).

The boundaries are considered far from the model so it is not affected by the analysis, and, on the other hand, the size of model and meshes are adjusted that the volume and time of the calculations will not be much (Fig. 3(a)). The ground stress of first set up is shown in Fig. 3(b). Based on this figure the stresses are uniform and first setup is correct.

Here, the constitutive model is plastic hardening with Mohr-Coulomb failure criterion. Plastic hardening model is a shear and volumetric hardening constitutive model for simulating of soil behavior. This model is described by the hyperbolic stress-strain relationship during drained axial compression, and the dependence of stiffness on stress is determined by a power. This model also includes shear and volumetric hardening rules and is based on the Mohr-Coulomb failure criterion (FLAC manual v.8). The

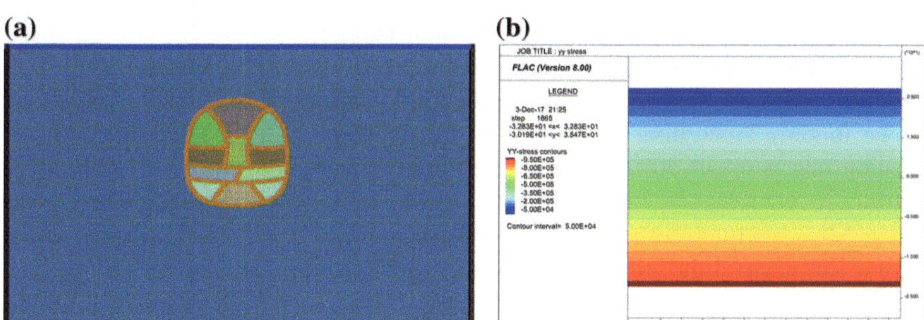

Fig. 3. (a) Finite difference model, (b) Fist set up

necessary parameters in this model include density, Modulus secant, modulus unloading, Poisson ratio, Cohesion and friction angle. These parameters are selected based on tests in laboratory.

After first setup, the first stage is excavated. It is relaxed up to desired value then the support system is installed. The amount of stress relaxation obtained is based on Hooke's relationships. These relationships are simplified for a circular tunnel under hydrostatic stress. Inductive deflection in the tunnel depends on the relative strength of the ground to the stress. To find the amount of relaxation, the excavation of the first drift is discussed. Based on the Hook's formula, the maximum radius of the plastic zone obtain 5.4 meters and, according to the longitudinal profiles provided by Vlachopoulos and Diederichs (2009), the displacement ratio to the maximum displacement is 5.5 and the radial deformation of tunnel is the 10 mm (Fig. 4).

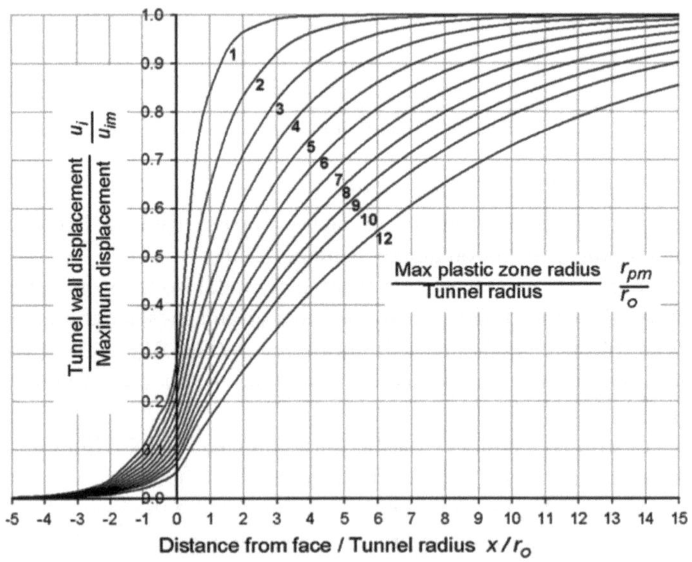

Fig. 4. Longitudinal displacement Profile [3]

Figure 5 shows the ground curve of the first excavation step (Fig. 4a). This plot depicts the displacement of the tunnel, without support, behind the face against the pressure of the ground relative to the in situ stress. According to this curve, as the tunnel face advances studied section, the internal pressure of the tunnel decreases and goes far to zero. According to this figure, for the displacement of 10 mm, the stress relaxation rate is about 30%. Because of simplified assumptions it isn't accurate and according to experiences, here it is increased to 40%. Then the supporting system, simulated by structural element, is installed.

To select excavation schemes 2 items are studied. They are shown in Fig. 6. The maximum settlement in both is almost the same (Fig. 7), but the amount of forces in (1) is smaller, so it is selected to implement.

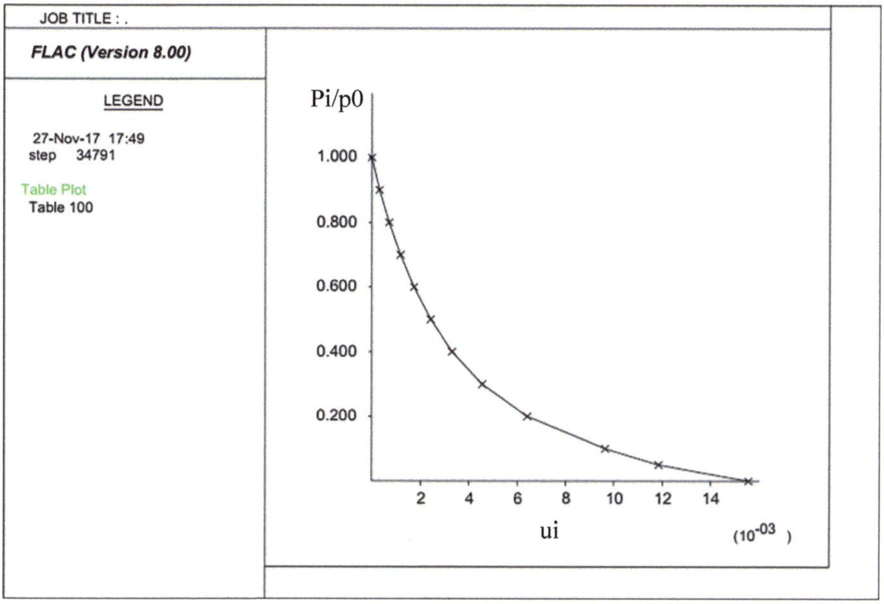

Fig. 5. Characteristic curve of stage 1

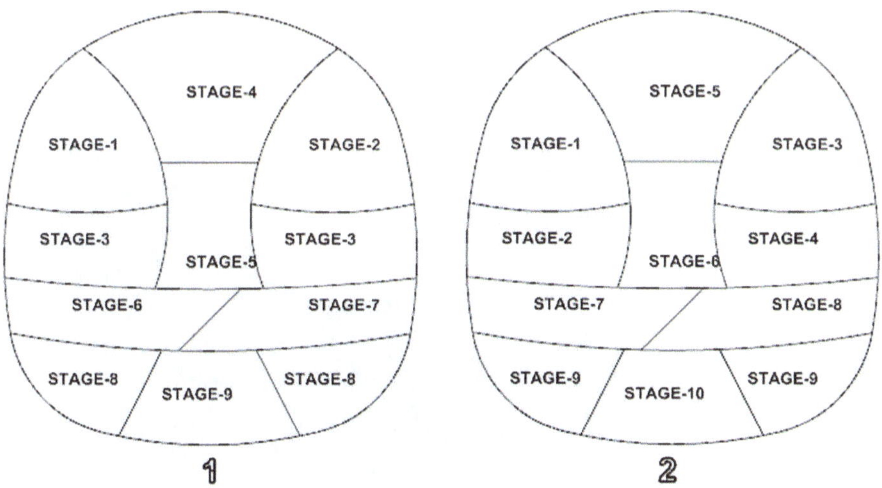

Fig. 6. The excavation schemes

The displacement contours and forces of the supporting elements are shown in Figs. 8, 9, 10 and 11. The maximum displacement in the tunnel crown is 4 cm because the overburden is 10 meters the displacement is a bit much. Also displacement is moved to the right of the station, because of the asymmetry of supporting system. The maximum force of shotcrete was 1300 kN and its maximum moment is 65 kN-m. The maximum force of the micropiles and nails were 154 kN and 140 kN, respectively (Fig. 12).

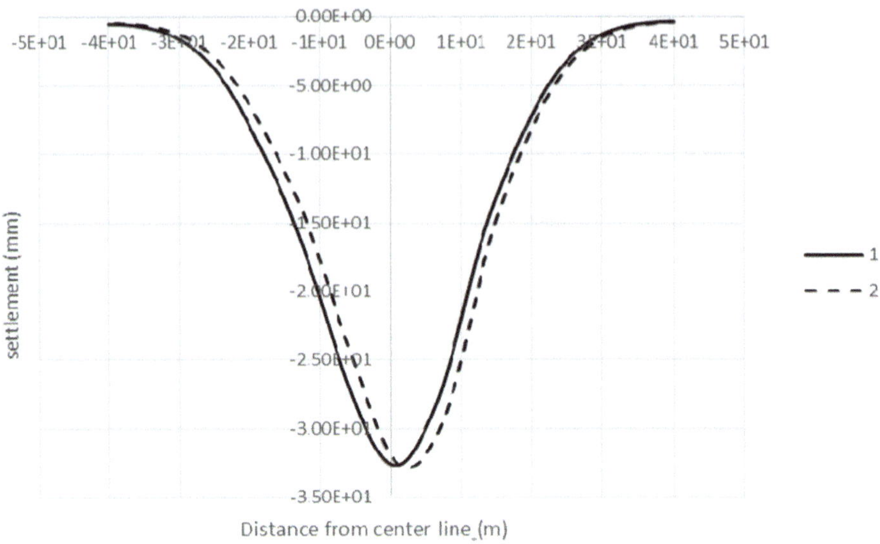

Fig. 7. Settlement of introduced excavation schemes (1) and (2)

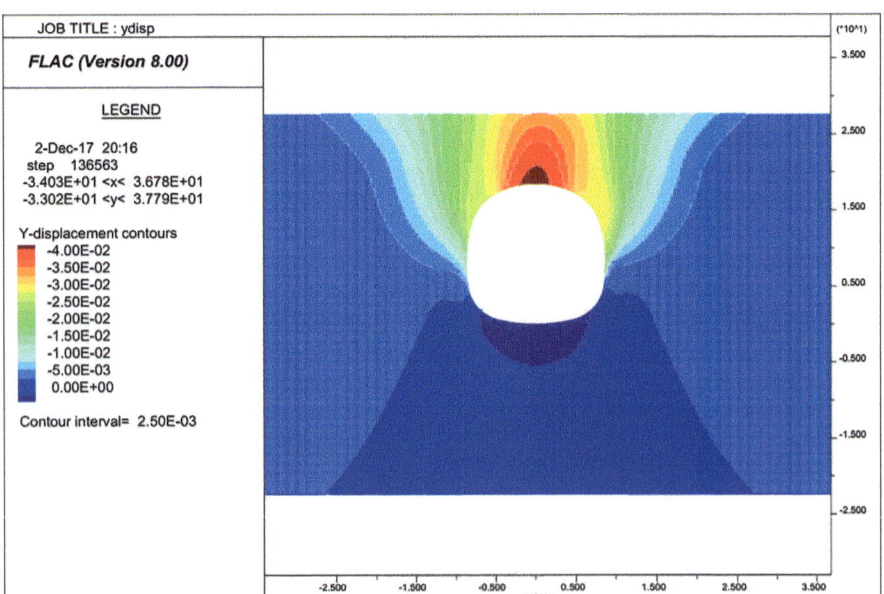

Fig. 8. Verical displacement contour

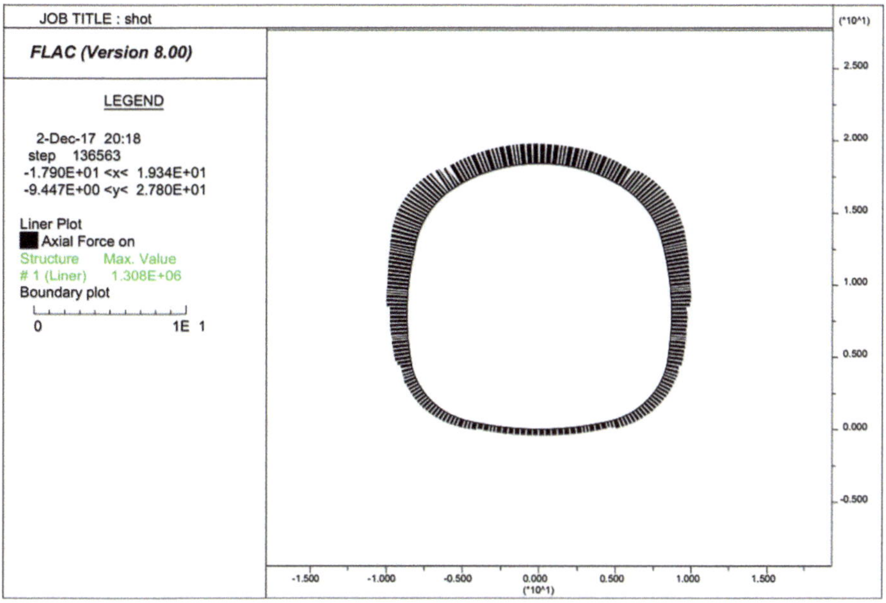

Fig. 9. Axial force in temporary lining

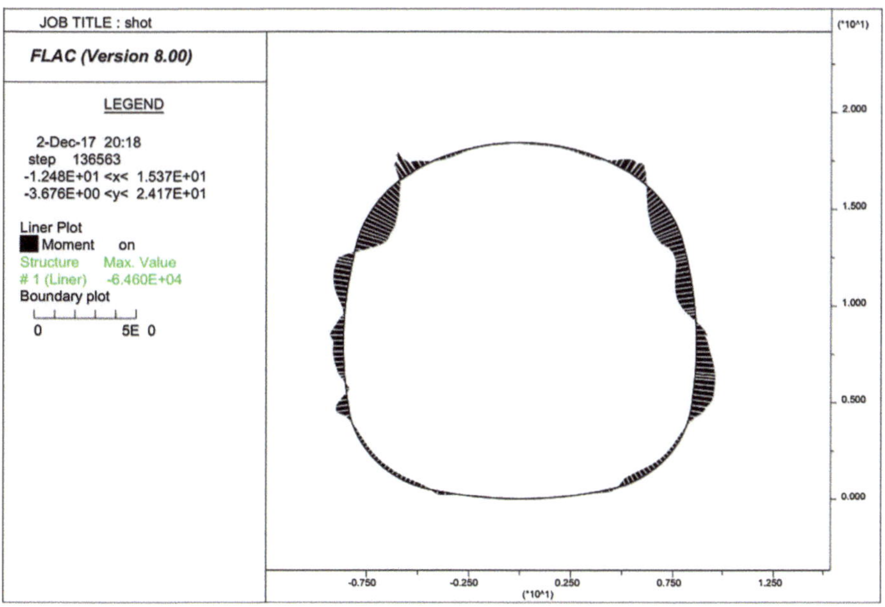

Fig. 10. Bending moment in temporary lining

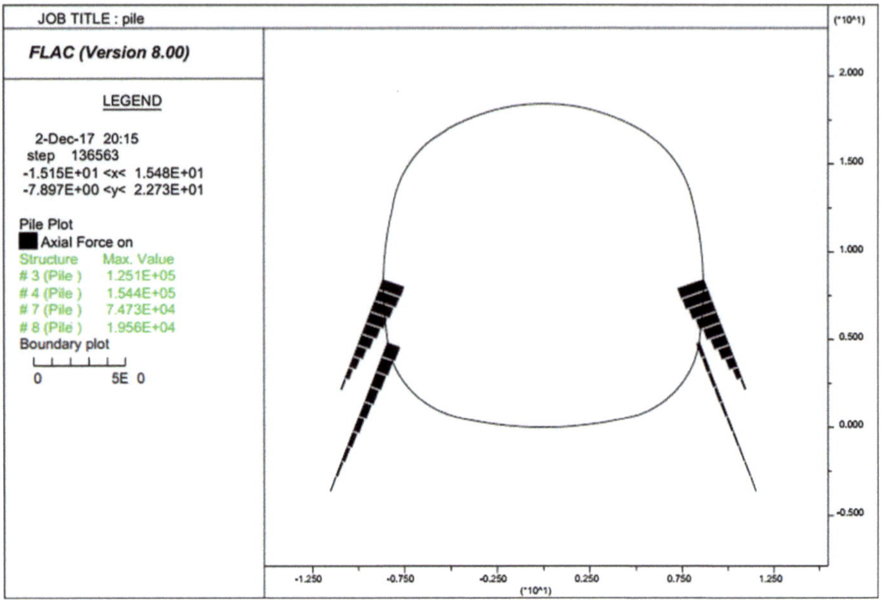

Fig. 11. Axial force in micropiles

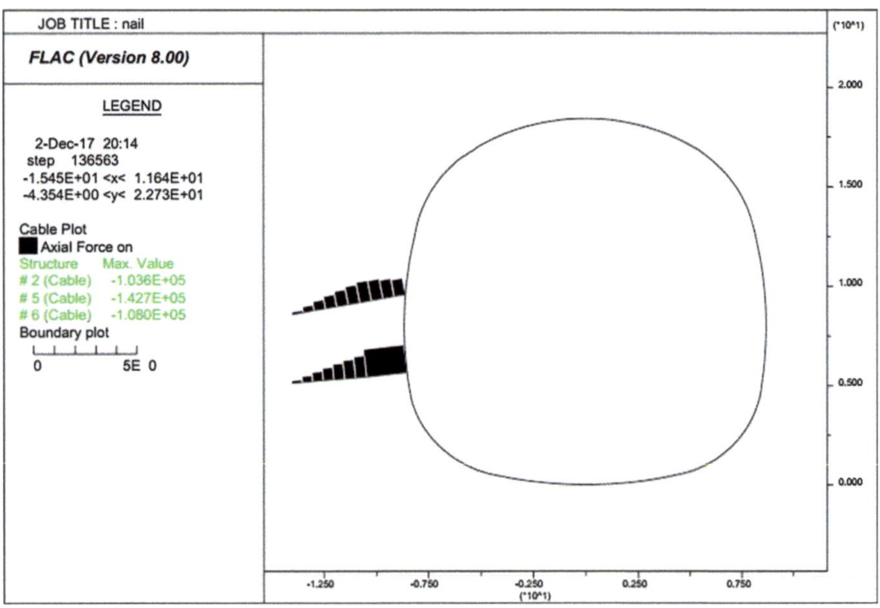

Fig. 12. Axial force in nails

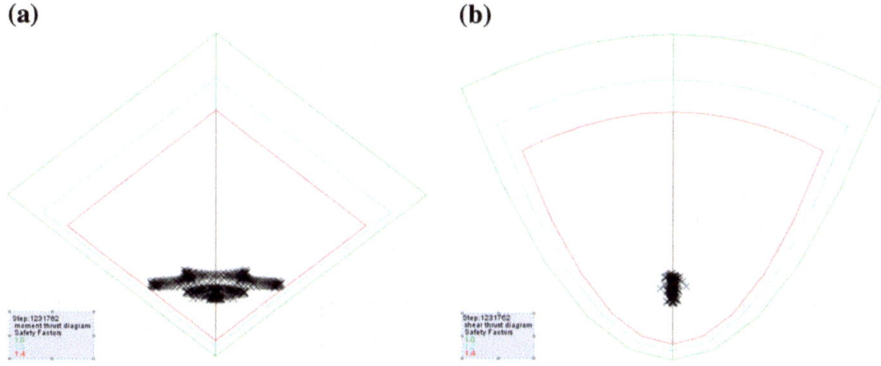

Fig. 13. Safety factor of (a): moment, (b): shear of station

The interaction of moment-thrust yield envelope and thrust-shear yield envelope of shotcrete and lattice are shown in Fig. 13(a, b). According to this figure the values are in allowable range, so the characteristic of temporary lining is acceptable.

5 Conclusions

In this study, a numerical method has been used to simulate a station with large section using underground excavation method. The convergence-confinement method (stress relief method) is used and the effect of the third dimension is considered. We select the construction scheme based on time and settlement. The results indicated that NATM method is a applicable for excavation large underground spaces such as a metro station.

Acknowledgments. The authors thanks the personals of the Metro Company for consulting.

References

Farias, M.M.D., Junior, A.H.M., Assis, A.P.D.: Displacement control in tunnels excavated by the NATM: 3-D numerical simulations. Tunn. Undergr. Space Technol. (2004). https://doi.org/10.1016/j.tust.2003.11.006

Hoek, E.: Rock-Support Interaction Analysis for Tunnels in Weak Rock Masses. Rocscience Inc, Canada

Itasca Consulting Group Inc, Sixth Edition (FLAC Version 8.0) February 2016, Minneapolis, Minnesota

Karakus, M., c, R.J.: Effects of different tunnel face advance excavation on the settlement by FEM. Tunn. Undergr. Space Technol. (2003). https://doi.org/10.1016/S0886-7798(03)00068-3

Kolymbas, D.: Tunnelling and Tunnel Mechanics. Springer, Berlin (2003). ISBN: 10 3-540-25196-0

Sharifzadeh, M., Kolivand, F., Ghorbani, M., Yasrobi, S.: Design of sequential excavation method for large span urban tunnels in soft ground – Niayesh tunnel. Tunn. Undergr. Space Technol. (2013). https://doi.org/10.1016/j.tust.2013.01.002

Vlachopoulos, N., Diederichs, M.S.: Improved longitudinal displacement profiles for convergence confinement analysis of deep tunnels. Rock Mech. Rock Eng. (2009). https://doi.org/10.1007/s00603-009-0176-4

Yu, C.W., Chern, J.C. Expert system for D&B tunnel construction. Undergr. Space (2007). ISBN: 978-0-415-40807-3

Classification of Seismic-Liquefaction Potential Using Friedman's Stochastic Gradient Boosting Based on the Cone Penetration Test Data

Jian Zhou(✉), Xin Chen, Mingzhen Wang, Enming Li, Hui Chen, and Xiuzhi Shi

School of Resources and Safety Engineering, Central South University, Lushan South Road, Changsha 410083, China
csujzhou@hotmail.com

Abstract. The analysis of liquefaction potential of soil due to an earthquake is a classical problem for civil and geotechnical engineers. In this paper, Friedman's stochastic gradient boosting (FSGB) method is introduced and investigated for the prediction of seismic liquefaction potential of soil based on the cone penetration test (CPT) data. The SGB models were developed and validated on a relatively large dataset comprising 226 field records of liquefaction performance and CPT measurements. The database contains the information about effective vertical stress, cone tip resistance, total vertical stress, sleeve friction ratio, depth of potentially liquefiable soil layer, earthquake magnitude and maximum horizontal ground surface acceleration. To find the most suitable model, several different combinations of above input parameters were tested to assess the usefulness of SGBs for liquefaction assessment using CPT data. SGBs are based on classification & regression trees with ensemble learning strategy and found to work well in comparison to artificial neural network and support vector machine models. The developed SGB provides a viable tool for practicing engineers to determine the liquefaction potential of soil.

Keywords: Liquefaction · Cone penetration test (CPT) · Stochastic gradient boosting

1 Introduction

Soil liquefaction triggered by earthquakes often causes damage to infrastructure (i.e., bridges, buildings, and lifelines) (Seed and Idriss 1971; Pal 2006; Goh and Goh 2007; Samui 2007; Oommen et al. 2010; Kohestani et al. 2015). Thus, it is essential to be able to estimate the potential of liquefaction occurrence and its effect in an earthquake-prone region. Also, the Cone Penetration Test (CPT) is being used increasingly in geotechnical investigations. Because the CPT has greater accuracy, easiness, repeatability and reliability compared with other field tests, and utility has found increasing use in earthquake engineering applications. Further, a number of liquefaction triggering curves based on CPT measurements are used in practice (e.g., Moss et al. 2006; Idriss

M. Barman et al. (eds.), *Transportation and Geotechniques: Materials, Sustainability and Climate*, Sustainable Civil Infrastructures,
https://doi.org/10.1007/978-3-319-95768-5_7

and Boulanger 2006). Thus, this paper focuses on estimation of the seismic liquefaction potential using the CPT data.

Over the past four decades, various approaches have been attempted to predicting seismic liquefaction potential of soil deposits, including empirical or semi-empirical methods, field methods, numerical methods, laboratory methods, analytical methods, soft computing and artificial intelligence based methods (*i.e.* Seed and Idriss 1971; Liao et al. 1988; Goh 1996; Lai et al. 2004, 2006; Idriss and Boulanger 2006; Moss et al. 2006; Pal 2006; Samui 2007; Oommen et al. 2010; Goh and Goh 2007; Xue and Yang 2013; Kohestani et al. 2015; Kaveh et al. 2016). However, few research works have been systematically performed to precisely predict and evaluate a level of safety of the structures. For example, empirical or semi-empirical approaches have employed with local monitoring data and are open to improvement because they are based on limited collected data. Furthermore, the estimation of reliable values of model input parameters is found to be an increasingly difficult task before applying sophisticated numerical methods. Especially, soft computing techniques such as artificial neural network (ANN), support vector machine (SVM), relevance vector machine (RVM), adaptive neuro fuzzy inference system (ANFIS) have been successfully implemented for evaluation liquefaction potential with better accuracy compared to available statistical methods in recent years. As an example, Pal (2006) investigates the potential of SVM - based classification approach to assess the liquefaction potential from actual standard penetration test (SPT) and CPT field data. Yazdi et al. (2012) combine a support vector data description and ANFIS to deal with the class imbalance problem in soil liquefaction data. Also, Xue and Yang (2013) employ an ANFIS and a data set of CPT-based records for evaluating soil liquefaction triggered by earthquakes. The ANN and ANFIS model are very time-consuming in the learning phase within which the membership functions and the rule base are constructed. A random forest algorithm is used by Kohestani et al. (2015) to predict the liquefaction potential of soil based on CPT data. Kaveh et al. (2016) employed the patient ruled-induction method (PRIM) to investigate the potential of liquefaction occurrence due to earthquake strikes in soils.In these studies, all the data were separated into training and testing sets. However, cross-validation (CV) process was not implemented and thus the accuracy of the predictive model is not fully understood but far from solving the problem of eismic-liquefaction potential completely. Hence, the issue of seismic-liquefaction potential prediction still pose considerable challenge for earthquake science.

Friedman's stochastic gradient boosting (FSGB), a relatively new algorithm, is a family of powerful supervised machine learning techniques that have shown promising results in terms of prediction performance, robustness and speed in a wide range of practical applications (Friedman 2001, 2002; Kuhn and Johnson 2013; Zhou et al. 2016a, b). To the best of our knowledge, however, seldom studies have reported the use of FSGB in classification type of seismic-liquefaction potential applications. It is therefore motivating to investigate the capability of FSGB in seismic-liquefaction potential prediction. The objective of this study is to investigate the feasibility of using the classification of the FSGB model to predict seismic-liquefaction potential in soil deposits. The theory and procedure of FSGB is briefly reviewed. The development of FSGB - based seismic-liquefaction potential performance model is presented using CPT data and the performance of a predictive model is also discussed.

2 Materials and Methods

2.1 Data Set and Predictor Variables

The database used in this work is based on the CPT case records compiled by Juang et al. (2003). The database consists of 226 cases (the numbers of non-liquefied and liquefied cases are 93 and 133, respectively), which are taken from CPT measurements at over 52 sites and field performance observations of six different earthquakes (San Fernando Valley earthquake in 1971, Heicheng earthquake in 1975, Imperial Valley earthquake in 1979 & 1981, Loma Prieta earthquake in 1989 and Chi-Chi earthquake in 1999). The six input variables used for liquefaction potential assessment are identified namely the cone tip resistance (qc, MPa), the sleeve friction ratio (Rf, %), the effective stress (S1, kPa), the total stress (S2, kPa), the maximum horizontal ground surface acceleration (amax) and the earthquake moment magnitude (Mw). The reader is referred to the references (Juang et al. 2003; Goh and Goh 2007) for details of these cases.

3 Materials and Methods

3.1 Details of Friedman's Gradient Boosting Algorithm

FSGB is a supervised machine learning technique for regression and classification problems, which produces a prediction model in the form of an ensemble of weak prediction models, typically decision trees (Friedman 2001). Using the connection between boosting and optimisation function approximation is viewed from the perspective of numerical optimisation in the function space, rather than in the parameter space (see Fig. 1). The first step estimates $\boldsymbol{a}_t$ by fitting a weak learner $h(\boldsymbol{x};\, \boldsymbol{a})$ to the negative gradient of the loss function using least-squares. In the second step, the optimal value of β_t is determined given $h(\boldsymbol{x};\, \boldsymbol{a}_t)$. The following is the algorithm for FSGB (Friedman 2001, 2002; Hastie et al. 2001; Lu et al. 2016; Zhou et al. 2016a, b):

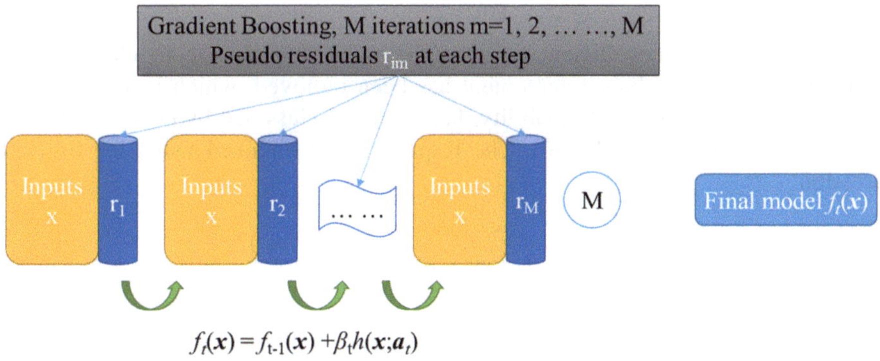

Fig. 1. Approach to FSGB

(1) Initialize $f_0(\boldsymbol{x})$ to be a constant, $f_0(x) = \arg\min_\beta \sum_{i=1}^{M} L(y_i, \beta), i = 1, 2, \cdots M$;

(2) For t = 1, 2, …, T, do the following:

- Compute the negative gradient as the working response $r_i = -[\frac{\partial L(y_i, f(x_i))}{\partial f(x_i))}]_{f(x)=f_{t-1}(x)}$;
- Fit a regression model to r_i by least-squares using the input x_i and get the estimate $\boldsymbol{a}_t$ of $\beta h(\boldsymbol{x};\boldsymbol{a})$;
- Get the estimate β_t by minimizing $L(y_i, f_{t-1}(\boldsymbol{x}_i) + \beta h(\boldsymbol{x}_i;\boldsymbol{a}_t))$;
- Update $f_t(\boldsymbol{x}) = f_{t-1}(\boldsymbol{x}) + \beta_t h(\boldsymbol{x};\boldsymbol{a}_t)$;

(3) Output $f(\boldsymbol{x}) = f_T(\boldsymbol{x})$.

Further detailed mathematical description over FSGB can be referred from Ref. (Friedman 2001; Hastie et al. 2001; Kuhn and Johnson 2013). To model the relationship between seismic-liquefaction potential and predictor variables, The FSGB model was implemented with the *gbm* R-package (Ridgeway 2007) within R environment software (Development Core Team 2017).

3.2 Evaluation of Classifier's Performance

The predictive power of FSGB algorithm on seismic-liquefaction potential data was evaluated by means of the classification accuracy rate and the Cohen's Kappa coefficient in this study (Hastie et al. 2001; Kuhn and Johnson 2013). Let $\boldsymbol{x}$ be an $Z \times Z$ contingency table or error matrix set out in rows and columns that express the number of sample plots (of which there are n) predicted to belong to one of Z classes relative to the true class label. Prediction accuracy, which is defined as the percentage of records that is correctly predicted by the model relative to the total number of records among the classification models, is a primary evaluation criterion. The classification accuracy rate can be obtained by

$$Accuracy = (\frac{1}{n}\sum\nolimits_{i=1}^{z} x_{ii}) \times 100\% \tag{1}$$

The Cohen's Kappa coefficient measures the proportion of correctly classified units after the probability of chance agreement has been removed, which is a robust index which takes into account the probability that a pixel is classified by chance (Kuhn and Johnson 2013; Zhou et al. 2016b). The Kappa can be obtained using the following expression

$$Kappa = \frac{n\sum\nolimits_{i=1}^{z} x_{ii} - \sum\nolimits_{i=1}^{z} (x_{i\cdot} \cdot x_{\cdot i})}{n^2 - \sum\nolimits_{i=1}^{z} (x_{i\cdot} \cdot x_{\cdot i})} \tag{2}$$

where x_{ii} is the cell count in the main diagonal, n is the number of examples, Z is the number of class values, and $x_{\cdot i}$, $x_{i\cdot}$ are the columns and rows total counts, respectively.

4 Results and Discussions

4.1 Descriptive Analysis

Figure 2 provides the relevant input parameters used to develop seismic-liquefaction potential prediction models range with their range, mean and outlets. The distributions of these variables and the relationship between maximum settlement and other input variables are demonstrated in the correlation matrix plot in Fig. 3, which can be seen the pairwise relationship between parameters with corresponding correlation coefficients and the marginal frequency distribution for each parameter. It can be also concluded that all parameters have no relatively good/meaningful correlation with each other and especially except with S_1 and S_2.

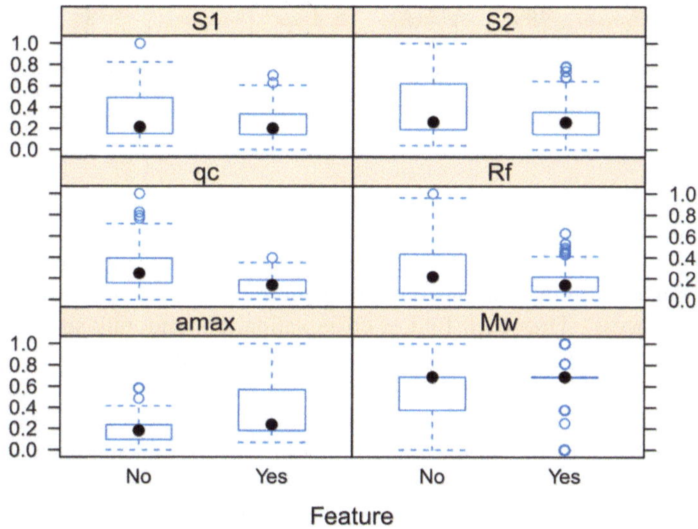

Fig. 2. Boxplot of variables for all data sets

4.2 FSGB Model Development and Validation

To measure the performance of the developed FSGB approach, in the present study, Data Set includes CPT-based 226 historical cases collected by Juang et al. (2003) and Goh and Goh (2007). This paper uses the above-mentioned methodology for the prediction of ground movements due to tunnelling. The predictive models are constructed using selected variables and Training set and applied to testing sets as shown in Fig. 3. Input parameters in the RF model are the factors influencing an evaluation target. In the RF package that we used, which is implemented in the R statistical environment (R Development Core 2017).

To estimate RSD, six indicators (qc, Rf, S1, S2, amax and Mw) were adopted as input variables and the measured seismic-liquefaction potential were taken as an output variable in the developed FSGB model. In order to determine reasonable and efficient

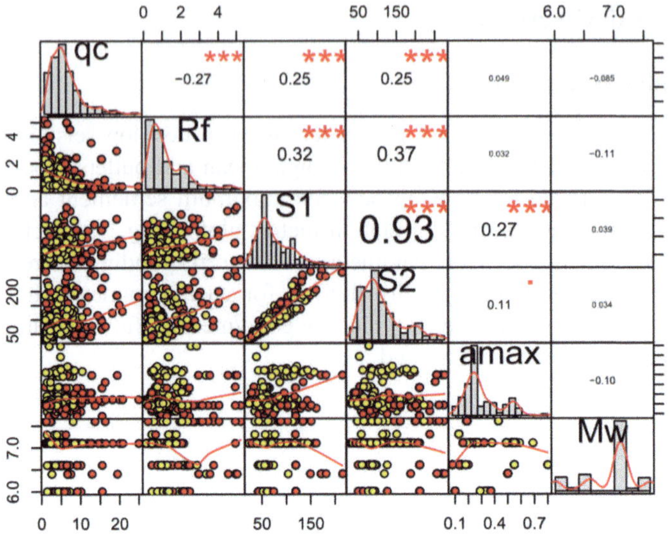

Fig. 3. The correlation matrix of variables for all data sets

the optimized hyper-parameters of FSGB, a 10-fold CV procedure (Kuhn and Johnson 2013; Zhou et al. 2015, 2016a, b) was implemented to determine the optimal parameter values based on the original training data set, with the original test removed completely from the CV process.. Thus each fold of cross validation randomly selects 10% of the data to serve as test data; then, the process fits a model on the remaining 90% of the data, as illustrated in Fig. 4. Classification tree (Kuhn 2008) was used as the single classifier in FSGB. Gradient boosting algorithms generally have four tweaking parameters which can be fine-tuned (Ridgeway 2007; Kuhn and Johnson 2013): n.trees (number of trees which is the number of gradient boosting iteration), interaction.depth (the complexity of the tree), shrinkage (learning rate) and n.minobsinnode (the minimum number of observations in trees' terminal nodes). If these parameters are not tuned correctly it may result in over-fitting. Thus it should be tuned aforementioned parameters for a good fit, and tuning parameters are set as follows: n.trees = {100, 200, 300, …, 900, 1000}; interaction.depth = {1, 3, 5, 7, 9, 11, 13, 15}; shrinkage = {0.001, 0.01, 0.05, 0.1}; n.minobsinnode = {5, 10, 20}. To find the most suitable model, moreover, several different combinations of above input parameters were tested to assess the usefulness of SGBs for liquefaction assessment using CPT data, i.e., model A: input variables = {qc, amax, Rf, S2, Mw and S1}; model A: input variables = {qc, amax, Rf, S2 and Mw }; and model C: input variables = {qc, amax and Rf}. Thus tuning parameters are considered optimized based on classification models that achieved the highest accuracy during the CV process.

To conduct the experiments, the dataset comprising of 226 cases is randomly splitted into two subsets. Thus the data are randomly split 80: 20 into training and test datasets. Accuracy was used to select the optimal model using the largest value, as shown in Fig. 5. In terms of model A: The final values used for the model were

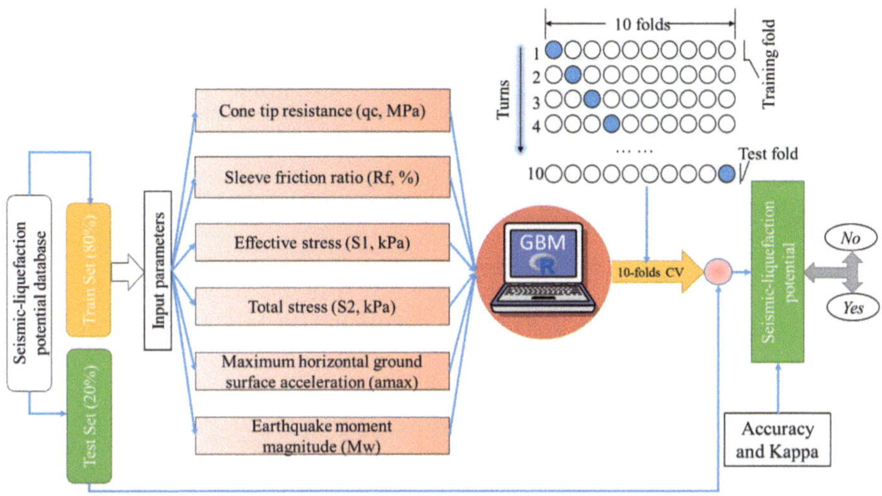

Fig. 4. Illustration of overall procedure for performance evaluation for seismic-liquefaction potential using FSGB methods

n.trees = 200, interaction.depth = 5, shrinkage = 0.1 and n.minobsinnode = 5. The OA and Kappa of the FSGB model are found to be 95.64% and 0.9108 for 182 sets of training data using 10-folds CV procedure, as can be seen from Fig. 5a. Similarly, the final values used for the model B (Fig. 5b) were n.trees = 200, interaction. depth = 9, shrinkage = 0.05 and n.minobsinnode = 5. The OA and Kappa of the FSGB model are found to be 95.11% and 0.8994, respectively; and the final values used for the model C (Fig. 5c) were n.trees = 200, interaction.depth = 3, shrinkage = 0.1 and n.minobsinnode = 10. The OA and Kappa of the FSGB model are found to be 90.04% and 0.7928, respectively.

To validate the predictive models based on the predicted and measured (real) values, 44 testing samples were validated by the optimized FSGB model. The results are presented in Table 1. producer's accuracy (PA) and user's accuracy (UA) (Congalton and Green 2009) for each class using the FSGB model is also presented in Table 1 based on the confusion matrices. The overall accuracy (OA) and Kappa of the FSGB model is found to be in the range of 86.36%–90.91% and 0.7179–0.8120 for 44 cases of testing data, respectively. The results are identical with field observations and the accuracy of this FSGB classification model is appropriate. The producer's and user's accuracy indicate that no liquefaction label is better classified than liquefaction one expect in model C.

4.3 Relative Importance of the Predictor Variables

The variable importance can be estimated accessible with the use of varImp() for FSGB models in *caret* package (Kuhn 2008). Variables are sorted by average importance across the classes. Figure 6 provides the result for the FSGB model using the generic function "*varImp*()" in the "*caret*" package and displays the relative variable

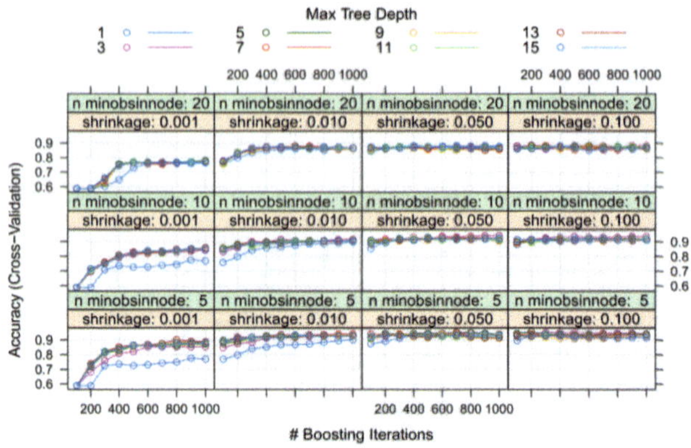

(a) Model A

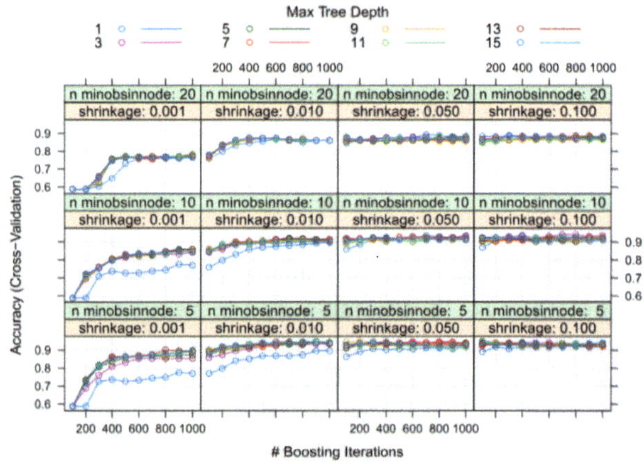

(b) Model B

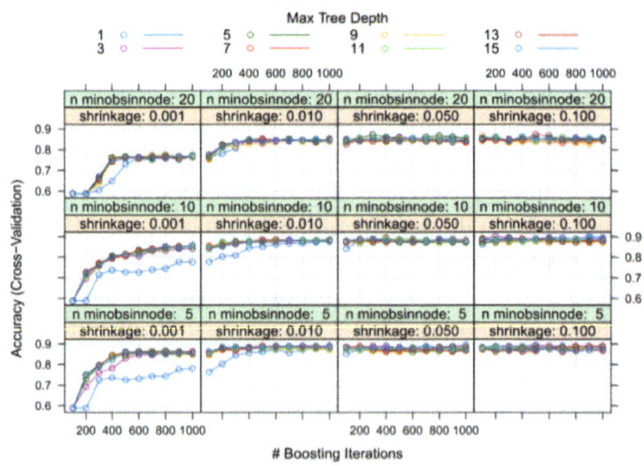

(c) Model C

Fig. 5. Plots of resampling profiles to examine the relationship between estimates of performance and tuning parameters for FSGB models with accuracy

Table 1. Confusion matrices and associated classifier performance for FSGB prediction model based on test data of seismic liquefaction potential

Model A				Model B				Model C			
	Reference				Observed				Observed		
Predicted	No	Yes	UA/%	Predicted	No	Yes	UA/%	Predicted	No	Yes	UA/%
No	**16**	2	88.89	No	**16**	3	84.21	No	**15**	3	83.33
Yes	2	**24**	76.92	Yes	2	**23**	80.00	Yes	3	**23**	88.46
PA/%	88.89	76.92		PA/%	88.89	88.46		PA/%	83.33	88.46	
Kappa = 81.20 OA = 90.91 [78.33, 97.47]				Kappa = 76.69 OA = 88.64 [75.44, 96.21]				Kappa = 71.79 OA = 86.36 [72.65, 94.83]			

The diagonal elements (correct decisions) are marked in bold importance for each of the six predictor variables. The sensitivity analysis on the indicators is carried out at the same time in order to discover the effects of variant indicators for estimation of these case instances. The *ROC* statistic is calculated as a relative measure of the variable importance, and the result is depicted in Fig. 6. Among all the variables, it is noteworthy that qc and a_{max} are the two most influential parameters among the indicators for prediction seismic liquefaction potential, followed by the indicator *Rf*, *S2*, *Mw* and *S1*.

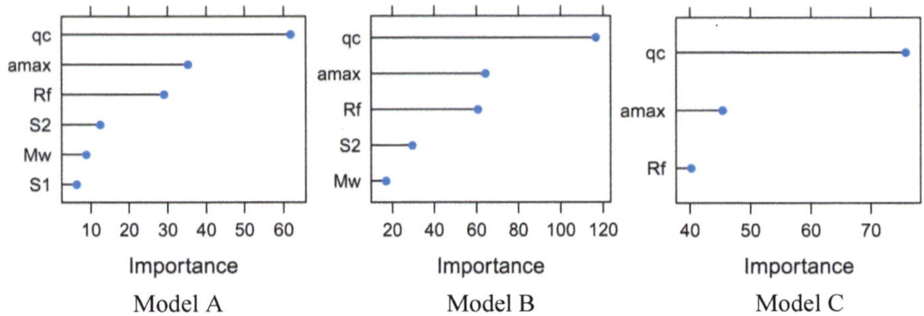

Fig. 6. Variable importance plot generated by the FSGB algorithm

5 Conclusions

Determination of seismic liquefaction potential of soil deposits is an imperative task in earthquake science. In the present study, a new approach is proposed for the prediction of seismic liquefaction potential of soil deposits incorporating the Friedman's stochastic gradient boosting. The applicability of FSGB for the prediction of ground movements is first checked with 226 case histories. Six different predictive variables that characterize soil and earthquake parameters are considered as input variables to identify sub-regions where liquefaction phenomena may occur. To examine the goodness of fit of the model, tuning parameters are considered optimized based on classification models that achieved the highest accuracy during the CV process. It's established that the FSGB has been successfully employed to investigate the potential of liquefaction occurrence due to earthquake strikes in soils. We also investigate which parameter is the most influential on the prediction of seismic liquefaction potential of soil deposits. This suggests that the FSGB can be effectively used to model the complex relationship between the liquefaction potential and different soil parameters, and also recommend its applicability to other geotechnical engineering failure issues.

Acknowledgments. The authors appreciate the support of the State Key Research Development Program of China (Grants 2016YFC0600706 and 2017YFC0602902) and the Sheng Hua Lie Ying Program of Central South University (Principle Investigator: Jian Zhou).

References

Friedman, J.H.: Stochastic gradient boosting. Comput. Stat. Data Anal. **38**(4), 367–378 (2002)

Friedman, J.H.: Greedy function approximation: a gradient boosting machine. Ann. Stat. 1189–1232 (2001)

Goh, A.T.: Neural-network modeling of CPT seismic liquefaction data. J. Geotech. Eng. **122**, 70–73 (1996)

Goh, A.T., Goh, S.: Support vector machines: their use in geotechnical engineering as illustrated using seismic liquefaction data. Comput. Geotech. **34**, 410–421 (2007). https://doi.org/10.1016/j.compgeo.2007.06.001

Hastie, T., Tibshirani, R., Jerome, J., Friedman, H.: The Elements of Statistical Learning, vol. 1. p. 339. Springer, New York, (2001)

Hu, J.L., Tang, X.W., Qiu, J.N.: Analysis of the influences of sampling bias and class imbalance on performances of probabilistic liquefaction models. Int. J. Geomech. 04016134 (2016)

Idriss, I.M., Boulanger, R.W.: Semi-empirical procedures for evaluating liquefaction potential during earthquakes. Soil Dyn. Earthq. Eng. **26**(2), 115–130 (2006)

Javadi, A.A., Ahangar-Asr, A., Johari, A., Faramarzi, A., Toll, D.: Modelling stress–strain and volume change behaviour of unsaturated soils using an evolutionary based data mining technique, an incremental approach. Eng. Appl. Artif. Intell. **25**, 926–933 (2012)

Juang, C.H., Yuan, H., Lee, D.-H., Lin, P.-S.: Simplified cone penetration test-based method for evaluating liquefaction resistance of soils. J. Geotech. Geoenviron. Eng. **129**, 66–80 (2003)

Kaveh, A., Hamze-Ziabari, S.M., Bakhshpoori, T.: Patient rule-induction method for liquefaction potential assessment based on CPT data. Bull. Eng. Geol. Environ. 1–17 (2016)

Kohestani, V.R., Hassanlourad, M., Ardakani, A.: Evaluation of liquefaction potential based on CPT data using random forest. Nat. Hazards **79**(2), 1079–1089 (2015)

Kuhn, M.: Building predictive models in R using the caret package. J. Stat. Softw. **28**(5), 1–26 (2008)

Kuhn, M., Johnson, K.: Applied Predictive Modeling. Springer, New York (2013)

Lai, S.Y., Hsu, S.C., Hsieh, M.J.: Discriminant model for evaluating soil liquefaction potential using cone penetration test data. J. Geotech. Geoenviron. Eng. **130**(12), 1271–1282 (2004)

Lai, S.Y., Chang, W.J., Lin, P.S.: Logistic regression model for evaluating soil liquefaction probability using CPT data. J. Geotech. Geoenviron. Eng. **132**(6), 694–704 (2006)

Liao, S.S., Veneziano, D., Whitman, R.V.: Regression models for evaluating liquefaction probability. J. Geotech. Eng. **114**, 389–411 (1988)

Lu, J., Lu, D., Zhang, X., Bi, Y., Cheng, K., Zheng, M., Luo, X.: Estimation of elimination half-lives of organic chemicals in humans using gradient boosting machine. Biochim. Biophys. Acta (BBA) **1860**(11), 2664–2671 (2016)

Moss, R.E., Seed, R.B., Kayen, R.E., Stewart, J.P., Der Kiureghian, A., Cetin, K.O.: CPT-based probabilistic and deterministic assessment of in situ seismic soil liquefaction potential. J. Geotech. Geoenviron. Eng. **132**(8), 1032–1051 (2006)

Oommen, T., Baise, L.G., Vogel, R.: Validation and application of empirical liquefaction models. J. Geotech. Geoenviron. Eng. **136**(12), 1618–1633 (2010)

Pal, M.: Support vector machines-based modelling of seismic liquefaction potential. Int. J. Numer. Anal. Meth. Geomech. **30**, 983–996 (2006)

R Core Team: R: A language and environment for statistical computing. R Foundation for Statistical Computing, Vienna, Austria. ISBN 3-900051-07-0, https://www.R-project.org/. Accessed 31 Mar 2017

Samui, P.: Seismic liquefaction potential assessment by using relevance vector machine. Earthq. Eng. Eng. Vib. **6**, 331–336 (2007)

Seed, H.B., Idriss, I.M.: Simplified procedure for evaluating soil liquefaction potential. J. Soil Mech. Found. Div. **97**, 1249–1273 (1971)

Xue, X., Yang, X.: Application of the adaptive neuro-fuzzy inference system for prediction of soil liquefaction. Nat. Hazards **67**(2), 901–917 (2013)

Yazdi, J.S., Kalantary, F., Yazdi, H.S.: Investigation on the effect of data imbalance on prediction of liquefaction. Int. J. Geomech. **13**(4), 463–466 (2012)

Zhou, J., Li, X.B., Mitri, H.S.: Comparative performance of six supervised learning methods for the development of models of pillar stability. Nat. Hazards **79**(1), 291–316 (2015)

Zhou, J., Shi, X.Z., Huang, R.D., Qiu, X.Y., Chen, C.: Feasibility of stochastic gradient boosting approach for predicting rockburst damage in burst-prone mines. Trans. Nonferrous Met. Soc. China **26**(7), 1938–1945 (2016a)

Zhou, J., Shi, X.Z., Li, X.B.: Utilizing gradient boosted machine for the prediction of damage to residential structures owing to blasting vibrations of open pit mining. J. Vib. Control **22**(19), 3986–3997 (2016b)

Variation of Lateral Soil Resistance Due to Two-Way Cyclic Loading

Myungjae Lee[1], Mintaek Yoo[2], Kyungtae Bae[3], Youseok Kim[3], and Heejung Youn[1(✉)]

[1] Department of Civil Engineering, Hongik University, Seoul, Republic of Korea
geotech@hongik.ac.kr
[2] Korea Railroad Research Institute, Uiwang, Gyeonggi-do, Republic of Korea
[3] Daewoo Institute of Construction Technology, Suwon, Gyeonggi-do, Republic of Korea

Abstract. In this study, the variation of lateral soil resistance of the monopile were investigated through centrifuge tests. The soil used in the experiment was Jumunjin standard sand, and the soil deposit was created with a relative density of 80%. The static load test was carried out to obtain the static lateral capacity, and the magnitudes of cyclic load were determined at 30, 50, 80 and 120% of the static capacity. A hundred cycles were applied to the pile head with a frequency of 0.125 Hz. The lateral soil resistance and pile displacement of the monopile subjected to cyclic loading were calculated through experimental data. At 2 and 5 m depth, the soil resistance increased with increasing number of cyclic loading, while the resistance decreased at 7 and 9 m depth.

1 Introduction

Researches on the lateral soil resistance of piles has been carried out mostly for onshore pile. Unlike onshore piles, offshore piles are heavily affected by environmental loads caused by wind, wave, and tide. The common feature of such loads is that they periodically load the offshore structure, which has a great influence on the stability of the structure itself as well as on the soils surrounding the piles. For this reason, there have been many studies on the effect of cyclic load on offshore piles.

Achmus et al. (2009) have numerically studied the long-term behavior of the foundation by performing 10^2–10^4 times of cyclic loading on the monopile. They adopted Stiffness Degradation Model for soil to simulate the accumulated displacement resulting from cyclic loading. Møller and Christiansen (2011) performed the static and cyclic lateral loading test of monopole in small scale and evaluated the conventional p-y curves for static as well as cyclic loading. Kim et al. (2015) adopted SWM (Strain Wedge Model) to evaluate the accumulated displacement at pile head. The input strains for 1–10^5 cycles were calculated considering the accumulated strains of the surrounding soil caused by cyclic loading, which is larger than the static loading value. In their study, the accumulated displacement was more affected by the magnitude of cyclic loads than the number of cyclic loads.

M. Barman et al. (eds.), *Transportation and Geotechniques: Materials, Sustainability and Climate*, Sustainable Civil Infrastructures,
https://doi.org/10.1007/978-3-319-95768-5_8

The p-y curves were practically used to analyze the lateral behavior of the pile. The pile in soil is regarded as the beam laterally supported by nonlinear spring, which is replaced by p-y curves. The concept of p-y curve was initially introduced by Winkler (1867), and many p-y curves have been proposed considering the soil type (sand or clay), loading type (static or cyclic), and the water condition (saturated or unsaturated). The conventional static p-y curves for clay and sand are the one by Matlock (1970), and by Cox et al. (1974), respectively. For cyclic loading, the p-y curves by API (1987) are popular for offshore industries, and those by Reese et al. (1974) and NCHRP (2001) are frequently used as well.

As the demanding for large diameter pile grows in the offshore industries, there was a doubt whether the current p-y curves are still valid for large diameter pile; and the p-y curves by API are to be used with caution (DNV 2011). One of the available options to investigate the behavior of large diameter pile with little cost is using the centrifuge model test. Li et al. (2010) evaluated the lateral behavior of large diameter pile under one-way loading, and it was found the pile head displacement increased with the number of cycles, and the pile secant stiffness was investigated. Garnier (2013) carried out numerous centrifuge tests on the monopile under cyclic loading up to 75,000 cycles, and the bending moment, pile head displacement, and the soil resistance were obtained. The test was performed with one-way loading, and the degraded cyclic p-y curves were proposed for beam analysis.

Numerous efforts have been made to evaluate offshore piles under lateral cyclic loads, but most studies have been limited to laboratory scale tests or full scale test but with small diameter. This study used saturated sand to satisfy offshore condition and adopted centrifuge tests to simulate large diameter piles. Lateral soil resistance due to cyclic loading was collected along the pile over time, and cyclic loads of four magnitudes were selected and used to investigate the lateral cyclic response of offshore pile.

2 Centrifuge Model Test

2.1 Test Set-up

Series of centrifuge tests were performed using laboratory facility of the DAEWOO Institute of Construction Technology. The facility has 3 m of arm length, and the soil box weight up to about 12 kN can be rotated at 100 times the gravity acceleration. Figure 1 shows the used soil boxes, which was made of aluminum alloy in three sides and bottom plate, and transparent plate was used at the front to monitor the testing setup. There are two sizes of soil box used in the test; one for static and the other for cyclic loading test. The soil box for cyclic loading test is wider than that for static loading test because the two-way loading system was mounted on the top of soil boxes. The soil box is 800 mm in length and 600 mm in height. The width is 200 mm for the static loading test and 400 mm for the cyclic loading test.

The soil box is mounted on a centrifuge facility and rotated until it reaches 94.2 times the gravity acceleration. When the desired gravity acceleration is reached, the length or diameter of the pile used in the model test can be converted to the full scale by a factor of 94.2 (Ko 1988).

Fig. 1. Soil boxes for (a) static loading test, and (b) cyclic loading test

The Jumunjin sand was used to create the sand deposit in the soil box. Table 1 provides the properties of the sand; the specific gravity was 2.65, the maximum and minimum dry unit weights were 16.6kN/m^3, and 13.3kN/m^3, respectively. Internal friction angle was measured to be 38° for 80% of relative density using direct shear test. Figure 2 presented the particle size distribution of the sand; the sand was classified into SP (poorly graded sand) according to the Unified Soil Classification System (ASTM D2487 2011). The effective grain size (D_e) of the sand was calculated to be 0.38 mm, and the pile diameter in the model test was 86.8 times the D_e, indicating the particle size has little effect on the centrifuge model test. In fact, Ovesen (1979) reported that the model test results would not be affected by particle size if the pile diameter is more than 30 times the effective grain size (D_e).

Table 1. Material properties of Jumunjin sand

USCS		SP
Dry unit weight (kN/m^3)	Maximum	16.6
	Minimum	13.3
Coefficient of uniformity (C_u)		1.68
Specific gravity (G_s)		2.65
Relative density (%)		80
Internal friction angle (°)		38

Sand was deposited using raining technique aiming at 80% relative density. The soils were prepared in dry condition, and water was supplied to the bottom of soil box through a pre-installed tube (Fig. 3). The water inflow was slow enough not to disturb the sand deposit nor trap the air in the void. The saturated sand deposit was left over for 1 day for full saturation before running test.

The pile is tubular made of aluminum with a 70 GPa of elastic modulus, 624 mm long, 35 mm in outer diameter, and 3 mm in thickness. The pile is embedded 424 mm into the Jumunjin sand. With the scale factor of 94.2, the pile dimension in the model test can be converted into 40 m in embedment depth, 60 m in length, 0.075 m in thickness and 3.3 m in outer diameter (Table 2).

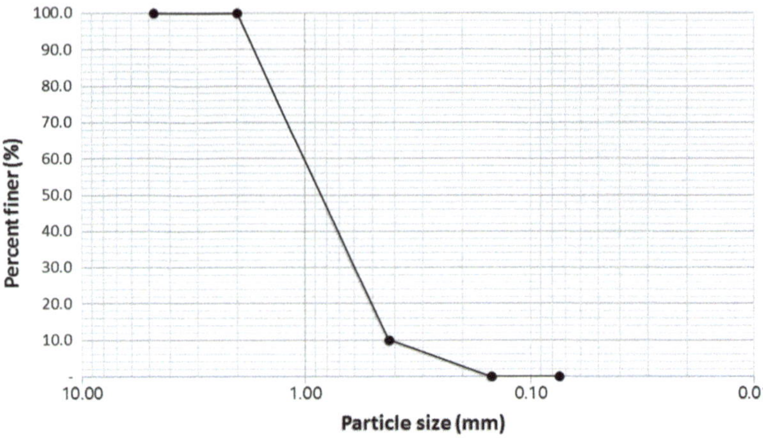

Fig. 2. Particle size distribution of Jumunjin sand

Fig. 3. Saturating tubes installed at the bottom of the soil box

Table 2. Pile dimensions in model scale and full scale

	Model pile	Full plie
Pile length	624 mm	60 m
Embedment depth	424 mm	40 m
Outer diameter	35 mm	3.3 m
Flexural rigidity (EI)	2.73 kN m^2	2.1×10^8 kN m^2

*Scale factor $\lambda = 94.2$

2.2 Static Loading Test

Figure 4 shows the schematic view of testing equipment for static loading test. Twelve strain gauges were attached to the pile along the depth in pair at 0, 2, 5, 7, 9 and 15 m from the ground level. Linear Variable Differential Transformer (LVDT) was installed to the pile head in order to measure the lateral movement, and a wire was connected from the pile head to the pullout motor with the load cell in the middle. The distance from the pile to the boundary was 382.5 mm, which is the 10.9 times the pile diameter. This satisfies the criteria to minimize the boundary effect of model test (Remaud 1999). The pile head was horizontally pulled out with the speed of 2.5 mm/min to failure in the static loading.

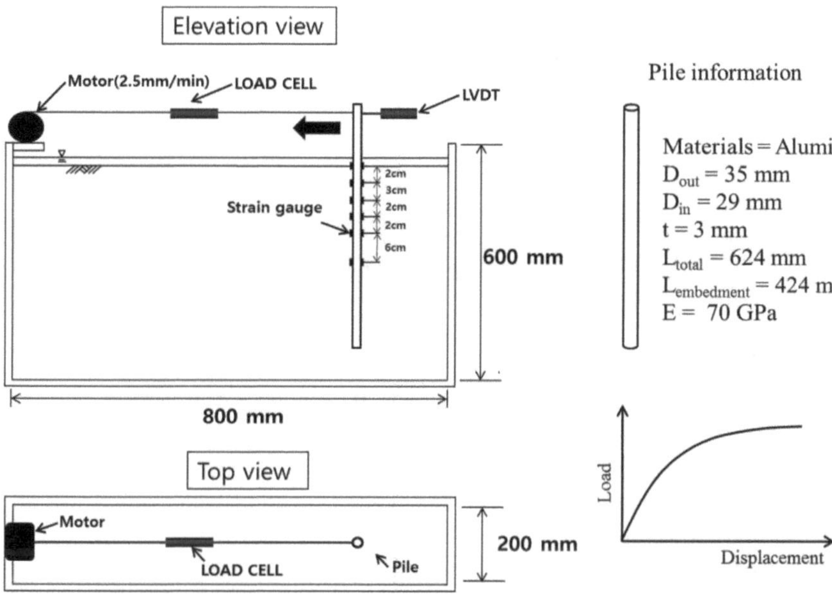

Fig. 4. Elevation and top view of static loading test

2.3 Cyclic Loading Test

Figure 5 shows the cyclic loading test setup used in this study. The pile with the identical dimensions of the static loading test was placed at the center of the soil box. Two LVDTs were installed at the pile head, and two load cells, one for each direction, were installed in the middle of each wire to measure the load in both directions. Accordingly, the lateral load-displacement curves were collected for each cycle at the pile head. Total of 16 strain gauges were installed in pair at 0, 2, 5, 7, 9, 15, 20 and 25 m depths.

Two-way cyclic loading was employed in the test, and the sinusoidal loading was generated through cam movement and the weights at each side. The cyclic loading device was slightly modified from the device developed for 1 g model test (Peng et al. 2006). The magnitudes of cyclic load were determined at 30, 50, 80 and 120% of the static capacity (1563 kN at the pile head). Table 3 tabulates the applied lateral load of five tests including four cyclic and one static loading test. The first symbol of each case

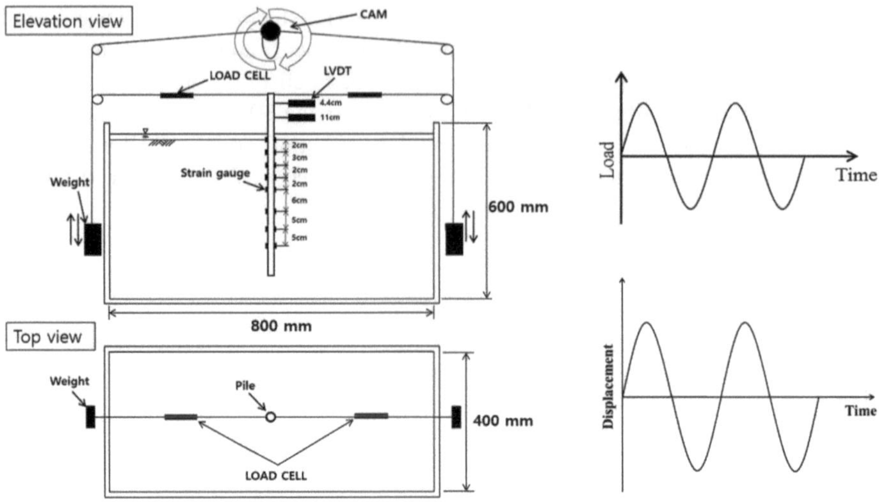

Fig. 5. Elevation and top view of cyclic loading test

Table 3. The magnitudes of cyclic loads and frequency

Case	Cyclic load (N)	Frequency (Hz)	Relative density (%)
S-80	N/A	N/A	80
C-80-569	569	0.125	
C-80-942	942		
C-80-1501	1501		
C-80-2256	2256		

indicates the loading type (S for static, C for cyclic), and the second number means the relative density, and the third number is the applied cyclic load before conversion to full scale value.

2.4 Lateral Soil Resistance and Displacement

The lateral soil resistance (p) as well as the lateral displacement (y) were calculated from the measured strains (ε) by calculating the moment distribution along the pile. The lateral soil resistance, p, is calculated by differentiating the moment distribution twice, and the displacement, y, is calculated by integrating it twice and divided by EI as shown in the Eq. (1). Cubic spline interpolation technique was used to derive moment distribution for the pile (Dou and Byrne, 1996).

$$\mathrm{p} = \frac{d^2}{dz^2}M(z),\ y = \iint \frac{M(z)}{EI}dz \tag{1}$$

where p is soil resistance (F/L), y is displacement of pile (L), EI is the flexural rigidity of the pile ($F \cdot L^2$), and M(z): moment distribution along the pile (F•L).

3 Test Results

Lateral static capacity was determined by load-displacement curve obtained from the static loading test. The test results from centrifuge test were converted into values in full scale with the scale factor of 94.2. Figure 6 provides the load-displacement curves at the pile head, and the static capacity was determined to be 1563 kN at 10% the pile diameter (Fleming et al. 1992).

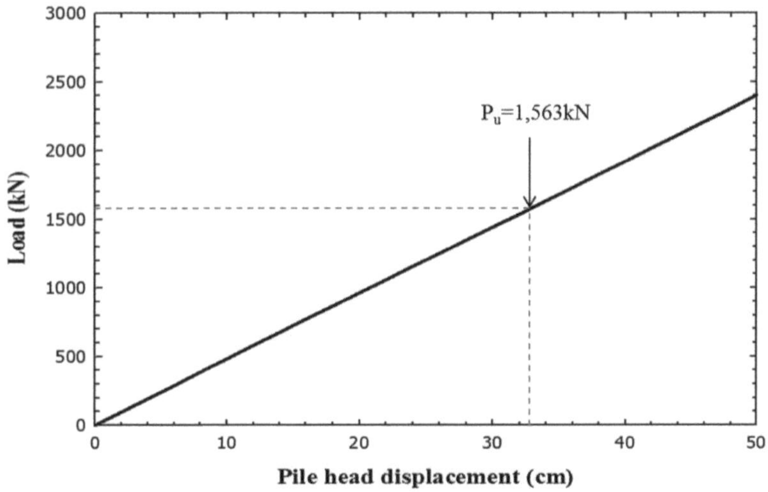

Fig. 6. Load—pile head displacement (at ground level) curve of static loading test

Figure 7 displays lateral soil resistance, p, and displacement, y, with time. Among the four cyclic loads, test results for the 569 N were selected for display. It is shown that the soil resistance gradually increases with time at relatively shallow depths (2 and 5 m) whereas it decreases with time at relatively deep depths (7 and 9 m). At shallow depths, cyclic loads seem to be sufficient enough to form gap between pile and soil, and the collapse of surrounding sand is likely to contribute to the densification around the pile, resulting in the increase in soil resistance. On the other hand, at deeper depths, the soils resistance decreases with number of cycles possibly because of the accumulation of pore water pressure resulting from cyclic loading. Similar trend was observed for the other three magnitudes of cyclic loads.

Figure 8 shows the experimental cyclic response in p-y plane at three different depths (2, 5, 7 and 9 m). The p-y curves were created at the 100th cycle of each cyclic loads and depths. The soil resistance appears to increase linearly with displacement with the small magnitude of cyclic loading, but for the 120% of the static capacity

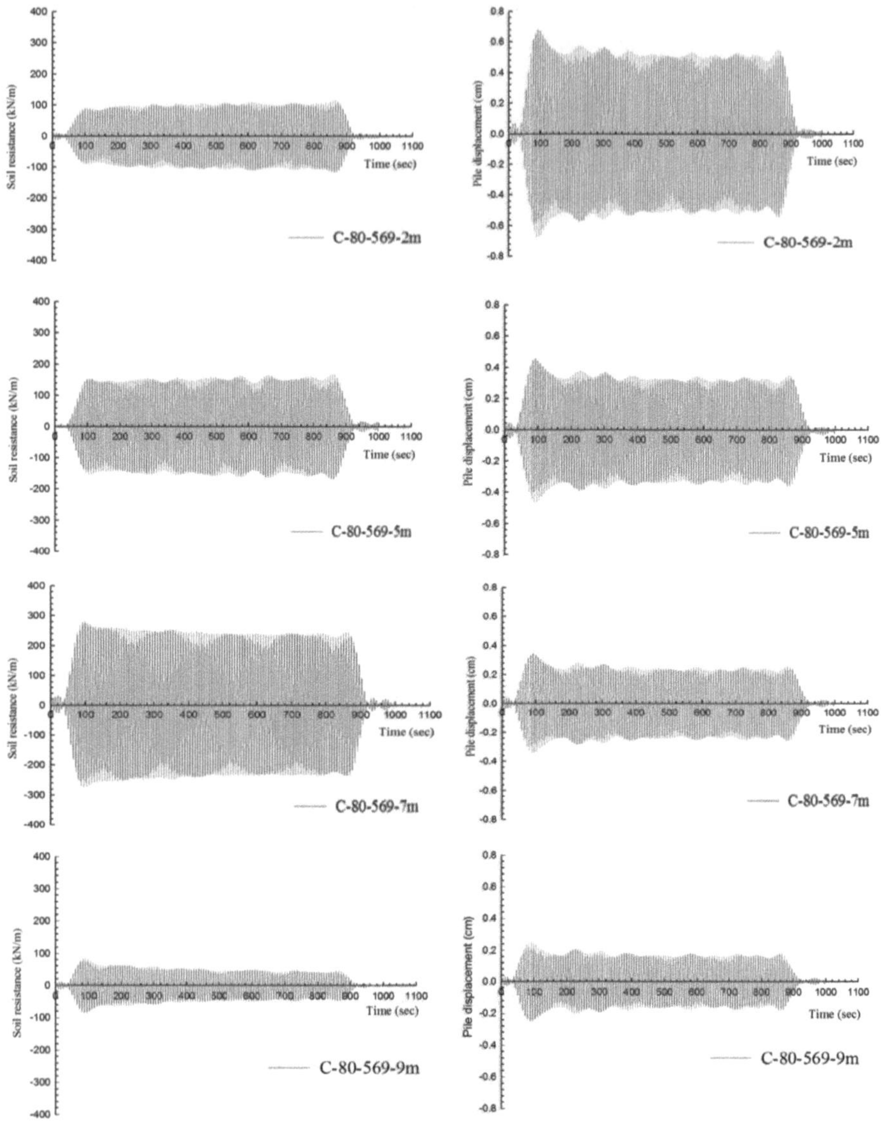

Fig. 7. Soil resistance and pile displacement with time for 569 N of cyclic load

(2256 kN), the soil resistance increases nonlinearly with the displacement. In addition, in was observed that the soil resistance as well as initial slope of p-y curve significantly increases with depth, indicating the pressure dependent behavior of sand.

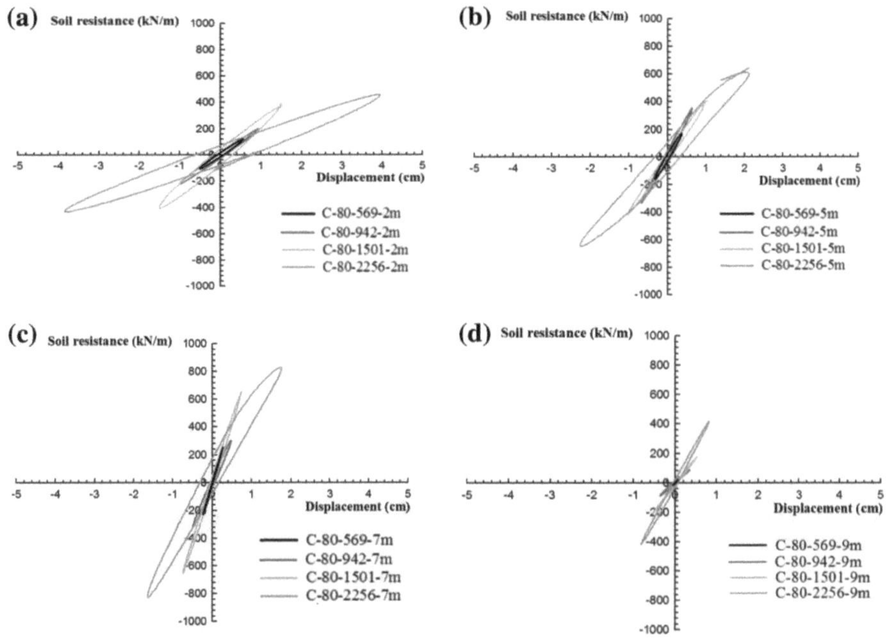

Fig. 8. Cyclic p-y curves at 100th cycle at depths of (a) 2 m, (b) 5 m, (c) 7 m, and (d) 9 m

4 Conclusions

In this study, the cyclic behavior of the monopile in saturated sand was investigated using centrifuge test. From the measured static capacity, four cyclic loads were determined and used to load the monopile in lateral direction. Two-way loading condition was used, and the soil resistance and pile displacement with time were collected along the pile. The following conclusions were drawn:

(1) At shallower depths (2 and 5 m), the soil resistance increases with the number of cycles, which is because of the densification of the surrounding sand under two-way loading condition. On the other hand, at depths greater than 7 m, the soil resistance decreases with the number of cycles. This is likely due to the increase in the pore water pressure resulting from cyclic loading.
(2) The lateral displacement decreases with the number of cycles along the pile. As the densification of the surrounding soil leads to larger soil resistance near surface, the resulting displacement for the same magnitude of cyclic loads decreases regardless of the depths.
(3) The cyclic p-y curves were created at the 100th cycle of each cyclic loads at four depths. The soil resistance as well as initial slope of p-y curve significantly increased with depth, indicating the pressure dependent behavior of sand.

Acknowledgments. This work was supported by National Research Foundation of Korea (NRF) funded by Ministry of Science, ICT & Future Planning (NRF-2016R1C1B2013478) and by the project titled as "Development of Substructure Systems for Offshore Wind Power in Shallow Sea Water (Less than 40 m)/2011T100201105" funded by the Ministry of Trade, Industry and Energy.

References

Achmus, M., et al.: Behavior of monopile foundations under cyclic lateral load. Comput. Geotech. **36**(5), 725–735 (2009). https://doi.org/10.1016/j.compgeo.2008.12.003

American Petroleum Institute (API).: Recommended Practice for Planning, Designing and constructing fixed offshore platforms. API Recommendation Practice 2A (RP 2A), 17th edn. (1987)

ASTM D2487.: Standard Practice for Classification of Soils for Engineering Purposes (Unified Soil Classification System), ASTM International, West Conshohocken, PA, (2011). https://doi.org/10.1520/D2487-11

Cox, W.R., et al.: Field testing of laterally loaded piles in sand. Offshore Technol. Conf. (1974). https://doi.org/10.4043/2079-MS

DNV, Det Norske Veritas.: DNV-RP-H103 Modelling and Analysis of Marine Operations (2011)

Dou, H., Byrne, P.M.: Dynamic response of single piles and soil pile interaction. Can. Geotech. J. **33**(1), 80–96 (1996). https://doi.org/10.1139/t96-025

Fleming, W.G., Weltman, A.J., Randolph, M.F., Elson, W.K.: Piling Engineering, 2nd edn. Wiley, New York (1992)

Garnier, J.: Advances in Lateral Cyclic Pile Design: Contribution of the SOLCYP project. In: Proceedings of TC 209 Workshop - Design for cyclic loading: piles and other foundations (ISSMGE), Paris, France (2013)

Kim, K., Nam, B.H., Youn, H.: Effect of Cyclic Loading on the lateral behavior of offshore monopiles using the strain wedge model. Math. Probl. Eng. **2015** (2015). https://doi.org/10.1155/2015/485319

Ko, H.Y.: Summary of the state-of-the-art in centrifuge model testing. In: Centrifuges in Soil Mechanics, pp. 11–18 (1988)

Li, Z., Haigh, S., Bolton, M.: Centrifuge modelling of mono-pile under cyclic lateral loads. 7th International Conference on Physical Modelling in Geotechnics, Zurich, vol. 2, pp. 965–970 (2010)

Matlock, H.: Correlations for design of laterally loaded piles in soft clay. In: Offshore Technology in Civil Engineering's Hall of Fame Papers from the Early Years, pp. 77–94 (1970)

Møller, I.F., Christiansen, T.: Laterally loaded monopile in dry and saturated sand—static and cyclic loading. Master Project, June (2011)

National Cooperative Highway Research Program. (NCHRP).: Static and dynamic lateral loading of pile groups, NCHRP Report 461. Transportation Research Board—National Research Council (2001)

Ovesen, N.K.: The scaling law relationship. In: Proceedings of the 7th European Conference on Soil Mechanics and Foundation Engineering, vol. 4, pp. 319–323 (1979)

Peng, J.R., et al.: A device to cyclic lateral loaded model piles. Geotech. Test. J. **29**(4), 1–7 (2006)

Reese, L.C., Cox, W.R., Koop, F.D.: Analysis of Laterally Loaded Piles in Sand. In: Offshore Technology in Civil Engineering Hall of Fame Papers from the Early Years, pp. 95–105 (1974)

Remaud, D.: Piles under Lateral Forces: Experimental Study of Piles Group. University of Nantes, France (1999)

Winkler, E.: Theory of elasticity and strength. Dominicus Prague (1867)

Subgrade Stiffness Effects on Mechanical Responses of Asphalt Pavement at Bridge Approach

Xinhong Yang[1], Yan Dong[2], Jiupeng Zhang[2](✉), and Hongbing Zhu[3]

[1] Shaanxi Provincial Expressway Construction Group Co., Ltd., Xi'an, Shaanxi, China
xinhongyang_sxgs@126.com

[2] School of Highway, Chang'an University, Xi'an, Shaanxi, China
{dongyan, zhjiupeng}@chd.edu.cn

[3] Liaoning Provincial College of Communications, Shenyang, Liaoning, China

Abstract. This paper investigates the effects of subgrade stiffness and stiffness transition on mechanical responses of asphalt pavement using multiple layered elastic analysis and three-dimensional finite element modeling. The surface deflections and internal stress distributions were analyzed by numerical simulation. The maximum shear stress was used as the indicator to determine the stiffness threshold for subgrade at bridge approach. It was found that the mechanical responses of asphalt pavement were very sensitive to the variation of subgrade stiffness. The stiffness of subgrade at bridge approach should be controlled more than 500 MPa, and the stiffness transition is better to follow an exponential function to ensure the uniform stress distribution of pavement structures. Considering engineering practice and cost factor, a reasonable stiffness range of 1000 MPa–2000 MPa was recommended for the subgrade at bridge approach.

Keywords: Bridge approach · Asphalt pavement · Subgrade · Stiffness transition · Mechanical responses

1 Introduction

Bridge approach plays a significant role in balancing structural stiffness and transiting settlement difference between bridge and pavement. Bridge abutment is made of rigid materials; thus the post-construction settlement is small. However, the embankment of bridge approach is usually made of flexible materials. The effect of consolidation settlement of foundation and embankment due to soil compression make the post-construction settlement even larger. The differential settlement between bridge abutment and embankment will cause two main distresses in bridge approach: pavement cracks and rough surface at bridge head.

Many treatment methods of bridge approach have been studied, such as installing approach slab, adding geotextile fabric, etc. Zaman et al. (1991) analyzed consolidation

M. Barman et al. (eds.), *Transportation and Geotechniques: Materials, Sustainability and Climate*, Sustainable Civil Infrastructures,
https://doi.org/10.1007/978-3-319-95768-5_9

settlement of embankment soil at bridge approach using nonlinear finite element method. Gao and Chen (1995) used the geotextile at the bridge abutment of high-grade highway. They analyzed the relationship between the backfill settlement at the back of abutment and the soil compaction degree, the thickness of geogrid fillings, and the length of geogrid. Zhang and Zhang (2003) analyzed the influences of different settlements and pavement structure parameters on the induced stress in the pavement using analytical equations. The effects of asphalt layer modulus and subbase modulus on the interface shear stress were especially analyzed. Feng et al. (2003) used the theory of elastic membrane to consider the settlement and deformation of different layers in the course of subgrade construction.

Han et al. (2005) studied the stress distribution of the deep mixed (DM) columns and the soft soil, and the distribution of tension along the reinforcement using numerical analysis. The results indicated that the maximum settlement developed at the base of embankment and the maximum tension occurred on the top of the DM columns. Bakeer et al. (2005) determined the factors affecting the long-term performance of bridge approach by examining a large number of pile-supported bridge approach slabs in southeastern Louisiana. Niu (2006) investigated vehicle bumping at bridge head on loess areas, showing that loess, sand, and gravel soil are the most common materials for abutment backfilling. Most of the backfilling projects did not set approach slabs, while the length discreteness of approach slab is larger somewhere setting approach slab, from 3 meters to 10 meters. Some of bridge approaches did not have differential settlement, or only had 1 cm differential settlement, but cracks were still observed on pavements. The damage of pavement structure was obvious at the end of approach slab. Xie et al. (2007) studied the active mechanism of geocell reinforcement and flexible approach slab for assimilating the foundation settlement and controlling the differential settlement at the bridge-subgrade transition section using three-dimensional numerical simulation.

Ge et al. (2007) compared the performance of bridge approaches using common backfill materials. Their study found that local materials should be used in abutment backfilling project after being improved and strengthened by new technologies (the integrated applying technology of expandable polystyrene interlayer and geogrid reinforcement). Gao et al. (2008) analyzed the deformation behavior of general fill material and expandable polystyrene (EPS) under self-gravity and vehicle dynamic loads, respectively. Sun (2010) presented two kinds of new methods of treatment through field study, the deep-seated concrete slab and expandable polystyrene (EPS) concrete replacement method. The results showed that the bridge approach treated by the deep-seated concrete slab and expandable polystyrene (EPS) concrete exhibited good effect on vibration reduction and isolation. Thiagarajan et al. (2012) presented bridge approach slabs designs for new and replacement slabs, including (a) cast-in-place (CIP) design for new construction and (b) novel, precast, prestressed (PCPS) slab designs for new construction and replacement of bridge approach slabs. They established finite element models to consider different slab lengths, thicknesses, and loss of support conditions to calculate maximum moments, deflections, and end slopes. Shi et al. (2013) built the three-dimensional finite element model of the bridge-embankment transition zone to explore the effect of anticipated future traffic on bridge approach. Robison and Luna (2004) discussed the differential settlement at the bridge abutment of two Missouri bridges using finite element method. They compared the measured displacements with simulation results. Coelho and Hicks (2015) designed transition zones at bridge approach by creating the gradual variation of stiffness. They assessed the performance of typical railway transition zone using numerical analysis and identified the critical train speed for the transition zone. Stark et al. (2016) reviewed the design of railroad track transition zone on bridge and monitored two well-performing bridge transitions using noninvasive accelerometers.

The aforementioned studies show that few studies have investigated the stiffness transition ratio, the minimum stiffness requirement, and the stiffness difference required at bridge approach. For these reasons, the backfill materials of many backfilling projects do not have enough stiffness, which results in cracking on pavements. Most of previous studies emphasized the importance of approach slab, however, it may form the second bumping at the end of approach slab and cause pavement cracking depending on slab length, thickness, and stiffness.

The aim of the paper is to analyze the sensitivity of pavement responses to subgrade stiffness and discuss the transition rules of subgrade stiffness at bridge approach. The ideal stiffness transition mode of bridge approach was proposed by analyzing the influence of stiffness variation on pavement stresses. The three-dimensional finite element model of bridge approach was established. The stiffness threshold of subgrade at back of abutment was given. The selection and quality of backfill material were strictly controlled according to the subgrade stiffness threshold. The length and the depth of approach slab can be determined by the stiffness threshold.

2 Sensitivity Analysis of Asphalt Pavement Responses to Subgrade Stiffness at Bridge Approach

Bridge approach is the structure connecting bridge and general roadway sections. The large difference of structure stiffness between bridge abutment and subgrade affects the stress state of pavement structure, which may result in deterioration of pavement performance. To balance the stiffness difference between bridge and pavement structure, the variation of stiffness at bridge approach need to be properly considered. According to current theories and assumptions, there is a transition rule of stiffness at bridge approach including the value of minimum stiffness and stiffness transition ratio. When the subgrade stiffness exceeds the minimum value, the influence of stiffness change on pavement structure can be neglected. When the variation of subgrade stiffness at bridge approach follows the transition rule, the stress state of pavement structure will experience the relatively uniform and continuous changes.

The road structure of bridge approach is not continuous in the base and subgrade, especially when the value of stiffness has sudden change and gradient. Therefore, the traditional road structure design principles cannot be directly used for bridge approach design. However, the sensitivity of pavement responses to subgrade stiffness can be used to determine the reasonable rules of subgrade stiffness change at bridge approach. Therefore, the stresses and strains in the pavement structure at different locations were calculated. The allowable stiffness variation at bridge approach was analyzed according to the changes of stresses and strains considering different base and subgrade stiffness values.

2.1 Pavement Structure and Parameters

Continuous construction method is widely accepted in the designs of bridge deck pavement, bridge approach pavement and general road pavement. In this method, the upper layer and the mid-surface of deck pavement and general pavement use the same structure and material to realize continuous state. Based on this principle, this paper established two kinds of typical pavement structures to study the mechanical behavior of pavement structure as the stiffness of bridge approach changes. With the goal of balancing the stresses and deformations of the deck pavement structure, bridge approach pavement structure and ordinary pavement structure, we determined the reasonable range of the stiffness at the bridge-subgrade transition section and established the law of the stiffness transition in bridge approach. The parameters used in this study are presented in Tables 1 and 2.

Table 1. Typical structure A—pavement with cement-treated base layer

Position / Horizon	Bridge abutment (bridge)	Bridge approach	General road
Surface layer	AC-10, 4cm E = 1500MPa,μ = 0.3		
Mid-surface	AC-16, 6cm E = 1200MPa, μ = 0.3		
Lower layer	——	AC-20, 8cm E = 1000MPa, μ = 0.3	
Base layer	——	Cement stabilized macadam, 40cm E = 1500MPa, μ = 0.25	
Subgrade	Cement concrete (bridge abutment) E = 300000MPa, μ = 0.2	Backfill materials E = 40~300000MPa, μ = 0.35~0.2	Soil matrix E = 40MPa μ = 0.35

2.2 Influence of Subgrade Stiffness on Deflection, Stress and Strain

This paper established two kinds of typical bridge approach structures, which has different types of base (semi-rigid base and flexible base respectively). The parameters of two bridge approach structures are presented in Tables 1 and 2. The overall stiffness of the subgrade in bridge approach changed in the range of 40–300000 MPa through changing elastic coefficient in software, so as to analyze the influence of the subgrade stiffness on the total deflection and the stress state of the pavement. The stress state of pavement structure was calculated using BISAR software, including surface deflection, stress and strain of surface layer, mid-surface and lower layer. Then three indexes (deflection, stress and strain) in three positions (bridge deck pavement, flexible base pavement and semi-rigid base pavement) were compared and analyzed. The calculated results are shown in Figs. 1–3.

Table 2. Typical structure B—pavement with asphalt-treated base layer

Horizon \ Position	Bridge abutment (bridge)	Bridge approach	General road
Surface layer	AC-10, 4cm E = 1500MPa,μ = 0.3		
Mid-surface	AC-16, 6cm E = 1200MPa, μ = 0.3		
Lower layer	——	AC-20, 8cm E = 1000MPa, μ = 0.3	
Base layer	——	ATB-25, 20cm E = 1200MPa, μ = 0.3	
Subbase	——	Graded crushed stone, 20cm E = 300MPa, μ = 0.35	
Subgrade	Cement concrete (bridge abutment) E = 300000MPa, μ = 0.2	Backfill materials E = 40~300000MPa μ = 0.35~0.2	Soil matrix E = 40MPa μ = 0.35

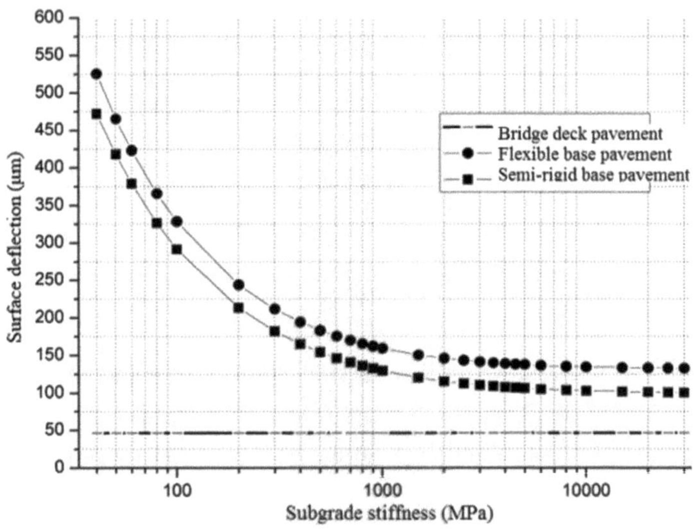

Fig. 1. Influence of subgrade stiffness on deflection

As can be seen from Fig. 1, the surface deflections of two pavement structures decrease with the increase of subgrade stiffness based on the calculation results, and the deflections show a plateau when the subgrade stiffness reaches 1000 MPa. When the subgrade stiffness is under 500 MPa, the surface deflection is more sensitive to the change of subgrade stiffness. The deflections of semi-rigid base pavement are always smaller than those of flexible base pavements, namely, the overall mechanical behavior of semi-rigid base pavement is closer to the upper surface stress state of bridge abutment.

As shown in Fig. 2, the surface layer, the mid-surface (lower layer) of bridge deck pavement and bridge approach are always in the compressive stress state. The compressive stress gradually decreases with the increase of the subgrade stiffness, and reaches stability at some point. This study assumes that when stiffness value changes in the range of 1000–2000 MPa, the stress condition of the bridge approach pavement is close to that of the bridge deck pavement.

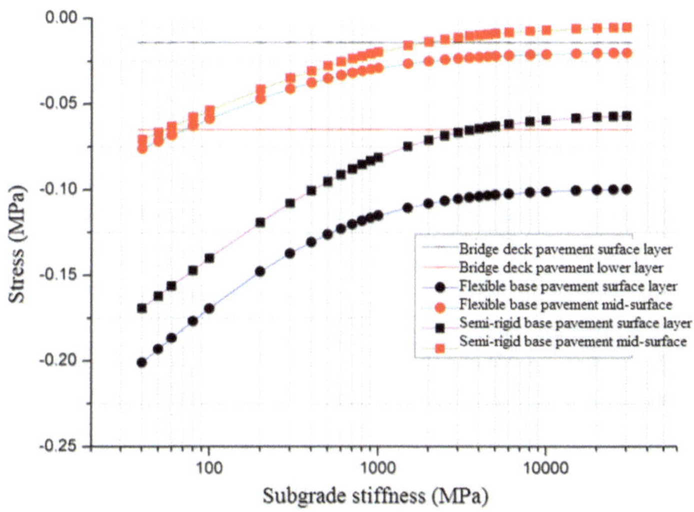

Fig. 2. Effect of subgrade stiffness on stress state of pavement structure

As shown in Fig. 3, the total strain state gap and total strain level between the surface layer and the mid-surface (lower layer) is very small, which is beneficial to maintain the continuous change of stress and strain in pavement structure. Compared with the flexible base pavement, the strain condition of semi-rigid pavement is closer to that of the bridge deck pavement.

2.3 Stiffness Transition Form Based on Surface Deflection

The subgrade stiffness of the general road is 40 MPa, and has the same pavement structure with bridge approach. It is found that when subgrade stiffness of bridge approach changes in the range of 40–200 MPa, the stress and strain states will change significantly. The idea of the stiffness transition is to have the stiffness reduced gradually from bridge abutment to general road through the range of bridge approach. To verify the relationship between the pavement stress state and the ratio of the stiffness transition, the author used surface deflection of semi-rigid base pavement (typical

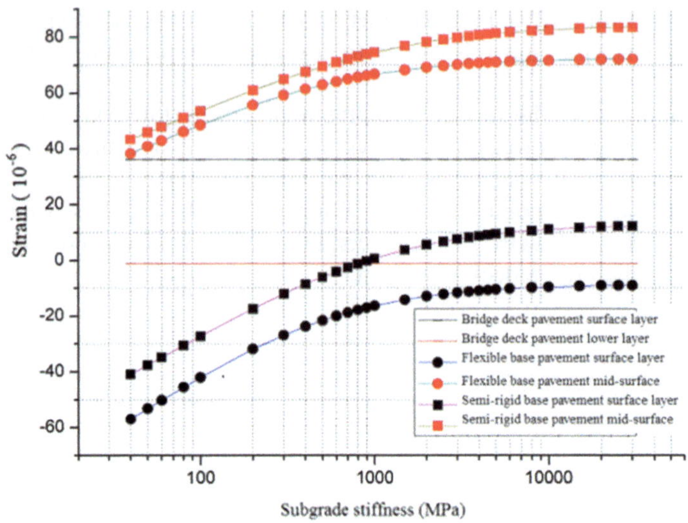

Fig. 3. Effect of subgrade stiffness on pavement strain condition

analysis structure A) as the control index, which is based on the results of the relationship between stiffness change and surface deflection. Firstly, logistic model was used to fit the relationship between deflection and stiffness variation. The logistic model is shown as Eq. (1):

$$y = A_2 + \frac{(A_1 - A_2)}{1 + \left(\frac{x}{x_0}\right)^p} \tag{1}$$

The fitting results and parameters are shown in Fig. 4. The relationship between surface deflection of the structure A and subgrade stiffness is shown as Eq. (2):

$$\omega = \frac{3147.64}{1 + \frac{E^{0.78}}{2.41}} + 95.76 \tag{2}$$

where, ω represents the surface deflection value, and E is the subgrade stiffness. This paper assumed a linear variation of the stiffness with the distance to the bridge abutment. When $L = 0$, E is the maximum value of subgrade stiffness 2000 MPa. When $L = L_0$, namely at the end of bridge approach which connects general road, $E = 40$ MPa, then the E is expressed as Eq. (3):

$$E = 2000 - \frac{1960}{L_0} \times L \tag{3}$$

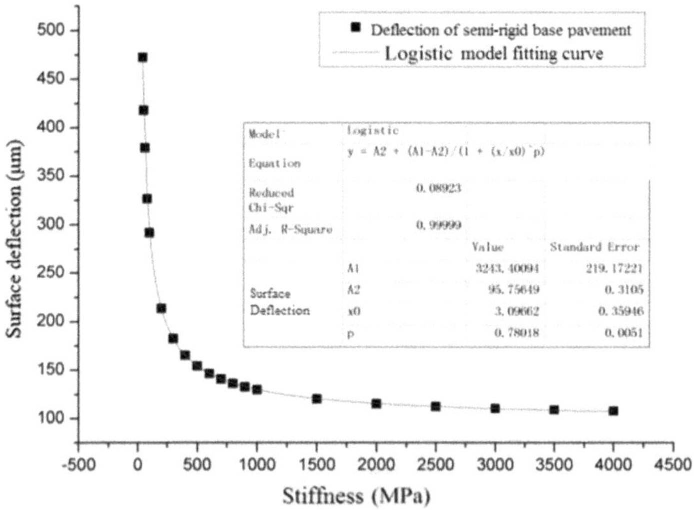

Fig. 4. Fitting of surface deflection

According to Eqs. (2) and (3), the relationship between surface deflection and the distance to bridge abutment can be represented by Eq. (4):

$$\omega = \frac{3147.64}{1 + \frac{\left(2000 - \frac{1960}{L_0} \times L\right)^{0.78}}{2.41}} + 95.76 \tag{4}$$

To visually express the relation, take the length of bridge approach L_0= 20 m, and the transition ration of stiffness is presented in Fig. 5. To make the stress and strain states of pavement structure transit uniformly at the same time, take surface deflection as the control index, and assume a linear variation of the deflection value with the increase of the distance to bridge abutment. Referring to the above approach, the relationship between stiffness and the distance to bridge abutment can be obtained from Eq. (5).

$$E = \sqrt[0.78]{\left(\frac{7585.81}{\frac{357}{L_0} \times L + 19.84} - 2.41\right)} \tag{5}$$

From Fig. 5, it can be easily observed that the change of the subgrade stiffness is non-linear when surface deflection changes linearly. When position is close to bridge abutment, the stiffness changes should accelerate because the subgrade stiffness is larger. When the subgrade stiffness reaches 200 MPa, the stress and strain states of pavement are more sensitive to the change of stiffness, and the position is gradually close to the general road. Thus, the subgrade stiffness is reduced at a relatively lower rate.

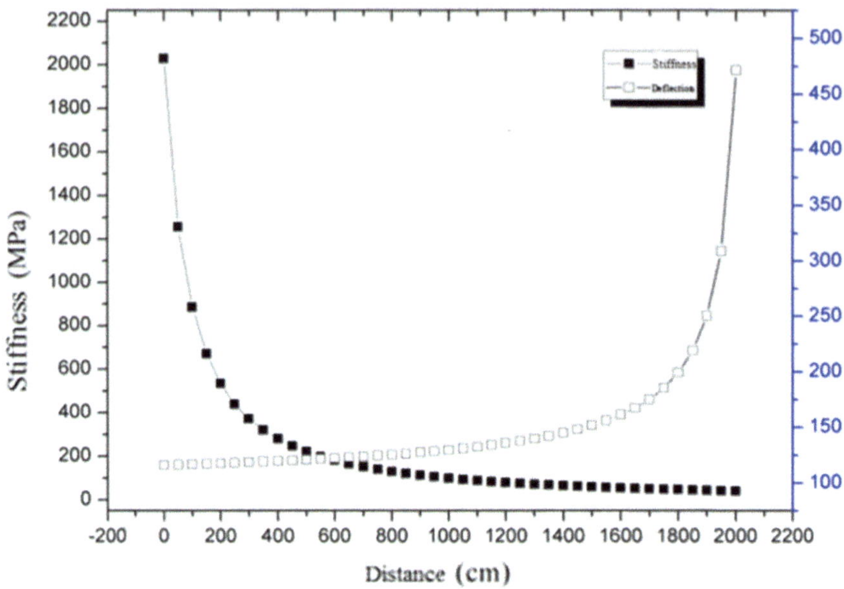

Fig. 5. Transition ratio of stiffness

3 Finite Element Model of Bridge Approach

Finite element analysis method provides a more efficient and lower cost way to evaluate the design concepts, so more and more people use finite element analysis method to solve structural analysis problems instead of doing model tests. ABAQUS is an engineering simulation software based on finite element method. ABAQUS has a set of powerful functions, which can solve all kinds of problems, from the relatively simple linear analysis to nonlinear simulation. ABAQUS has a very rich cell library, and it can simulate any actual shape; ABAQUS also has a wealth of material model library, which can simulate the nature of most typical engineering materials.

3.1 Geometric Model

In order to study the adaptability of pavement structure to stiffness difference at the back of abutment, the subgrade stiffness threshold was analyzed. Stiffness threshold ensures that the pavement structure is not destroyed by the stiffness difference. Meanwhile, this paper explored the influence of the stiffness transition mode on pavement structure. The three-dimensional (3-D) geometric model of pavement structure and embankment were established using ABAQUS 6.13. The illustration of 3-D finite element model is shown in Fig. 6.

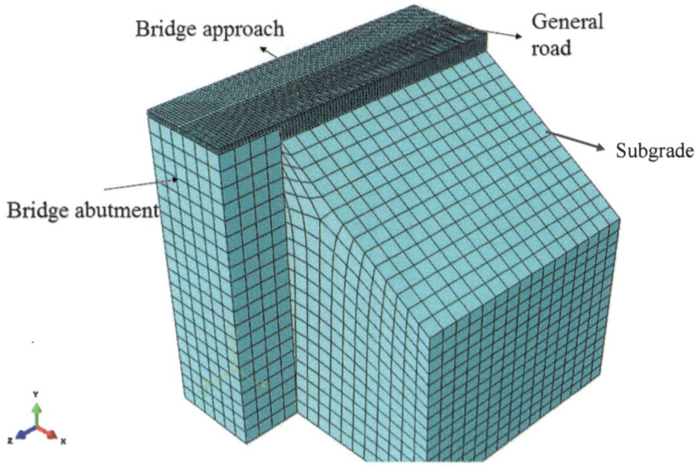

Fig. 6. Finite element model and meshes

This study considered the influence of model size, mesh generation and the boundary conditions on the calculated results and calculation efficiency. The larger the model size is, the more work will be done. In a similar way, the more dense the mesh is, the more work will be done, which will affect the work efficiency. Otherwise, if the size is too small or the mesh is rough, the accuracy of the calculation will be affected. The boundary conditions have a great influence on the calculation results, and it is very important to choose the correct boundary conditions.

So, in this paper, the model width of bridge abutment and pavement is 4 meters, and bridge abutment is 10 meters high. The surface layer adopts the "standard" structure, in which the thicknesses of surface layer, intermediate layer, and lower layer are 4 cm, 6 cm and 8 cm, respectively. The thickness of base course is 40 cm. The subgrade is 9.6 m high, 4 m width on the top surface, 1:1.5 slope settings. The thickness of the above structure layers and the slope of the subgrade slope are selected according to standard requirements (JTG D50-2017), and conform to the conventional expressway design proposal. Meanwhile, pavement and subgrade, bridge abutment are connected by contact (tie pairs). Subgrade and bridge abutment use friction contact, and the friction coefficient is 0.2. When the friction is included in model, the asymmetric phase is added in ABAQUS/Standard equation set. If the friction coefficient is less than 0.2, the value and influence of the asymmetric phase are very small, and the calculation effect is better (Zhuang 2005).

3.2 Simplified Loading Model

Wang et al. (2014) found that the distribution of contact stresses was affected by vehicle maneuvering and surface friction at the tire-pavement interface. According to the existing research results, the wheel load is presented in Fig. 7. The dual wheel load is simplified as rectangular loading with uniform pressure. The size of loading area is

0.22 m in length and 0.16 m in width, the distance between double rectangular centers is 0.32 m (Huang 1998). The load value is determined by Eq. (6), and the value is 0.7 MPa (JTG D50-2017).

Fig. 7. Simplified model of wheel loading

$$p = \frac{P}{S} = \frac{100\ \text{kN}}{4 \times 220\ \text{mm} \times 160\ \text{mm}} = 0.7\ \text{MPa} \tag{6}$$

3.3 Material Parameters

The structure of each layer is assumed as a linear elastic material with elastic modulus and Poisson ratio. The main research in this paper is the influence of subgrade stiffness change on pavement responses. The overall mechanical behavior of semi-rigid base pavement is close to the upper surface stress state of bridge abutment. Therefore, the semi-rigid pavement structure was selected as calculation model. The structure combination and material parameters are shown in Table 3.

Table 3. Material parameters in FE model

Structure layer	Materials and thickness	Elastic modulus	Poisson ratio
Surface layer	Asphalt concrete AC-10 4 cm	E = 1500 MPa	μ = 0.3
Mid-surface	Asphalt concrete AC-16 6 cm	E = 1200 MPa	μ = 0.3
Lower layer	Asphalt concrete AC-20 8 cm	E = 1000 MPa	μ = 0.3
Semi-rigid base	Cement stabilized crushed stone 40 cm	E = 1500 MPa	μ = 0.3
Subgrade	–	E = 3000–10000 MPa	μ = 0.2
		E = 10000–5000 MPa	μ = 0.25
		E = 5000–1000 MPa	μ = 0.3
		E = 1000–40 MPa	μ = 0.35
Bridge abutment	Cement concrete	E = 3000 MPa	μ = 0.2

4 Fem Simulation Results and Discussion

4.1 Stress Calculation Indices

The stiffness is discontinuous along the road in bridge abutment and bridge approach, which leads to the rearrangement and concentration of stress at the back of abutment. The subgrade stiffness and strength of bridge approach are obviously higher than that of general road, although both of them adopt the same structure and materials. Therefore, there is little value for determining the stiffness threshold of back filling behind abutment through the current stress calculation indices. By comparing stress of

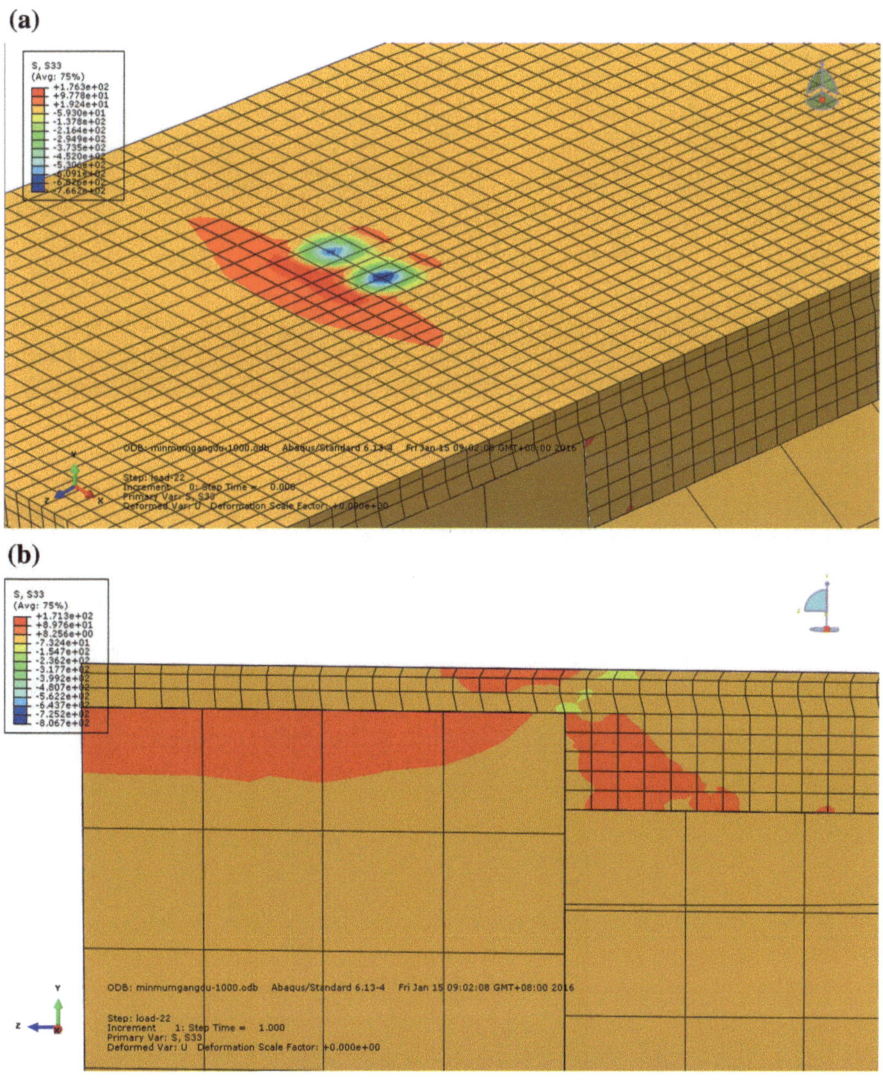

Fig. 8. **a** Longitudinal tensile stress of pavement surface. **b** Longitudinal tensile stress at the bottom of layer. **c** Vertical shear stress at the bottom of layer. **d** Maximum shear stress

different layers and locations under different subgrade stiffness, it is found that the structure discontinuous of bridge approach have a great influence on the stress indices of longitudinal and vertical. It is found that the primary influenced regions are located in the road surface and the bottom of lower layer in back filling behind abutment (Fig. 8).

(c)

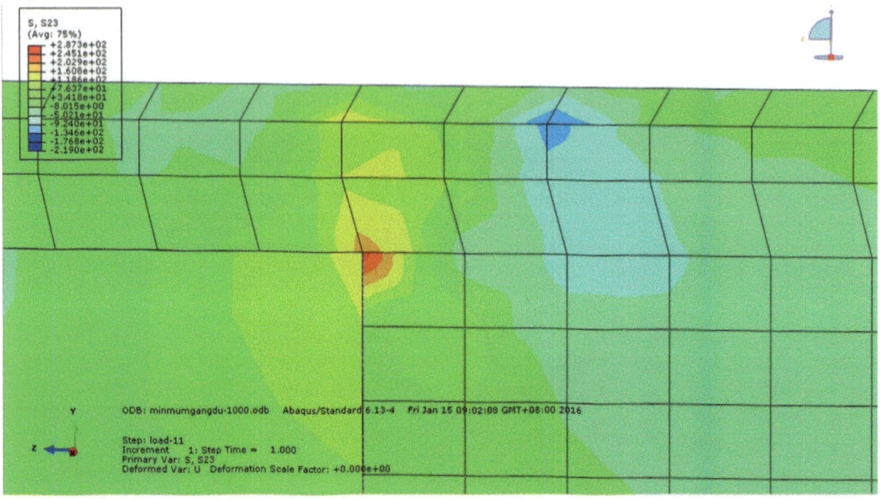

(d)

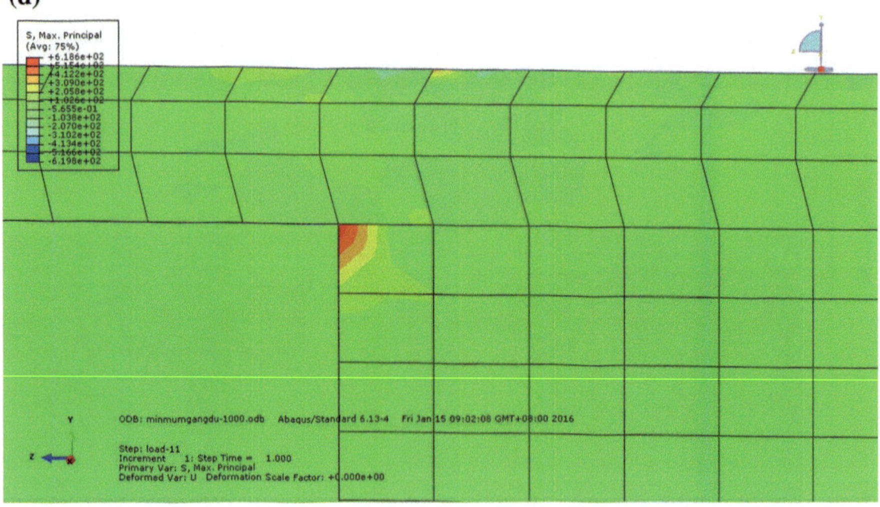

Fig. 8. (*continued*)

Based on the overall analysis of the data, four indices are selected to measure the influence of stiffness difference on stress state of pavement: longitudinal tensile stress of road surface on back of abutment S33, longitudinal tensile stress at the bottom of pavement on back of abutment S33, vertical shear stress at the pavement bottom on back of abutment S23, and the maximum shear stress τ_{max}.

4.2 The Position of Loading and Stress Calculation

Different loading positions are selected according to the needs of force analysis when establishing three-dimensional finite element model. Partition and refine pavement unit, then the specific meaning and position of loading, calculation points are shown in Table 4.

Table 4. Positions of loading and calculation points

Parameters

Loading position	Distance to bridge abutment (cm)
Position 1	-11
Position 2	0
Position 3	11
Position 4	22
Position 5	33
Position 6	44
Position 7	55
Position 8	350
Position 9	450
Position 10	550

Diagrams

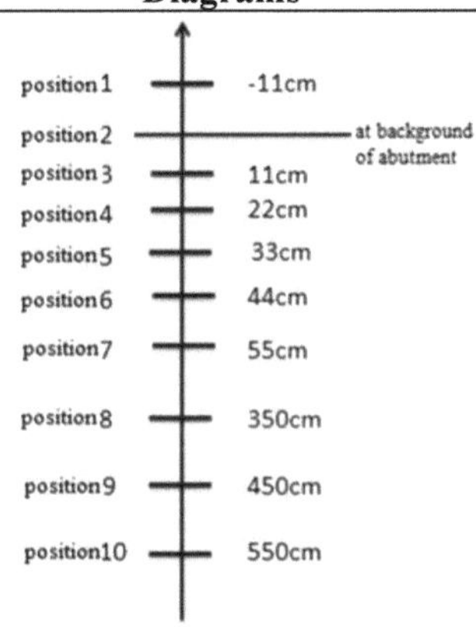

Calculation points	The distance of the double circle load center (cm)
Position 1	40
Position 2	32
Position 3	24
Position 4	16
Position 5	8
Position 6	0
Position 7	8
Position 8	24
Position 9	35.5

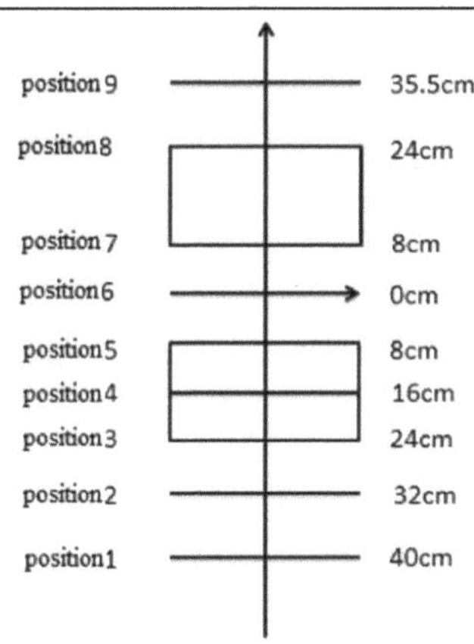

4.3 The Influence of Stiffness Change on Stress State of Pavement

The above analysis identified calculation indices of pavement stress, loading position and calculation point position at back of abutment. The pavement stress state was obtained when stiffness changes in the range of 5000–100 MPa. The calculation results are shown in Table 5 and Fig. 9.

Table 5. The stress state of pavement under different stiffness

Stiffness (MPa)	The tensile stress of road surface (kPa)	The tensile stress at the bottom of pavement (kPa)	The shear stress at the bottom of pavement (kPa)	The maximum shear stress of pavement (kPa)
5000	157	203	261	285
4000	158	204	263	288
3000	160	206	267	293
2000	164	209	274	301
1500	167	211	279	308
1000	171	215	287	319
900	173	216	290	322
800	174	217	292	325
700	176	218	295	329
600	178	220	299	334
500	181	221	303	339
400	184	224	308	345
300	189	226	314	353
200	195	229	323	363
100	207	233	335	378

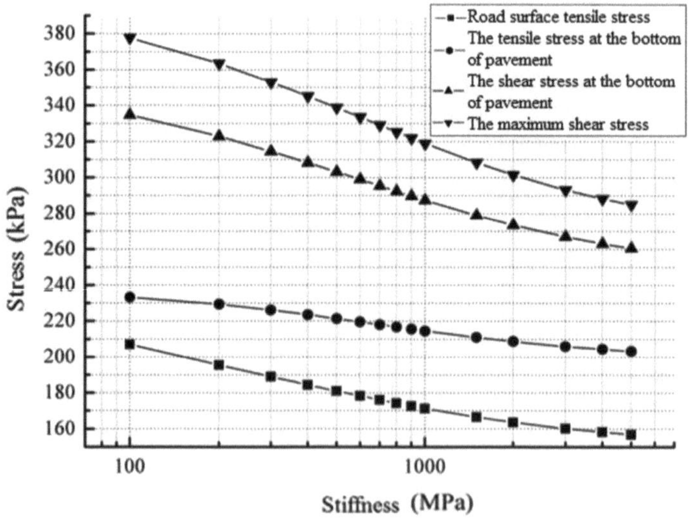

Fig. 9. The stress states of pavement under different subgrade stiffness values

As shown in Fig. 9, the shear stress changes most when subgrade stiffness changes. The bending tensile stress at the bottom of pavement is less sensitive to the change of stiffness. The bending tensile stress is lower as a design index of general road, which is much lower than the tensile splitting strength of common asphalt mixture 600–1000 kPa. Road surface tensile stress is much more sensitive to the change of stiffness than the tensile stress at the bottom of pavement, while the most sensitive one is the maximum shear stress (increasing 93 kPa). The vertical shear stress at the bottom of pavement and the maximum shear stress have similar variation law and occurred in the same position and condition. Thus, it is reasonable to use the maximum shear stress to measure the shear effect of pavement structure, which is caused by the difference of the stiffness and the loading effect. From the view of stress value, the longitudinal tensile stress of road surface and the longitudinal tensile stress at the bottom of pavement are both smaller, which will not cause structural damage of the pavement. So they cannot be used as the stress analysis indexes of pavement structure. In summary, the shear strength of the pavement should be satisfied when carrying out the design of pavement structure and material.

4.4 Stiffness Threshold

Generally, the maximum shear stress of pavement structure is in the range of 200–270 kPa. As showed in this study, the maximum shear stress of the pavement is 285 kPa when the subgrade backfill stiffness of the bridge approach reaches 5000 MPa. When the stiffness value is reduced to 500 MPa, the maximum shear stress will reach to 340 kPa, which accounts for approximately 1.2–1.7 times of the maximum shear stress of general road structure. Currently, the pavement rutting is due to the lack of the ample shear resistance in asphalt mixture. Zhang (2012) held that the shear resistance index was 385 kPa ignoring horizontal braking force, according to the stress analysis of large longitudinal slopes. Therefore, there is a shear stress control index that makes the fracture, fatigue damage and excessive plastic deformation disappear. Meanwhile, the strength standard also makes the most of the pavement materials can be easier to achieve.

Limited to the achieved extent of current theories and practices, this paper recommended to take the 1.5 times of the maximum shear stress in general pavement structure as the control index of the maximum shear stress. According to the calculation result statistics, Sun et al. (2013) found that when road withstands different combination forms and different forms of loading, the shear stress under double circular loads has little difference, and the average value is 224 kPa. In this paper, it is recommended that the maximum shear stress is controlled under 340 kPa, the corresponding minimum stiffness index is 500 MPa.

5 Conclusions

(1) The sensitivity of pavement responses to the change of subgrade stiffness was analyzed. It is concluded that when the subgrade stiffness is lower than 500 MPa, the stress state of pavement structure is sensitive to the change of subgrade

stiffness. However, when subgrade stiffness is higher than 2000 MPa, the effect of subgrade stiffness change on the stress state of pavement structure is less significant.

(2) According to the sensitivities of the pavement structure to subgrade stiffness, it is observed that surface deflection changes linearly when the subgrade stiffness varies exponentially.

(3) Finite element models were developed to analyze pavement response at bridge approach. Four response indicators were used to measure the influence of the different stiffness on the stress state of pavement. The results suggested that the vertical shear stress and the maximum shear stress can be used to determine the subgrade stiffness threshold at the back of abutment.

(4) By analyzing the maximum shear stress of pavement structure at general road, the maximum shear stress control criterion is 340 kPa based on the average value of the maximum shear stress and the 1.5 times the stress reserve. The threshold of the stiffness of the subgrade is 500 MPa. The reasonable range of subgrade stiffness should be within the range of 1000–2000 MPa considering engineering practice and costs.

Acknowledgments. This research was supported by the China Postdoctoral Science Foundation [grant number 2017M620434], the Department of Science & Technology of Shaanxi Province [grant number 2016KJXX-69, 2016ZDJC-24, 2017KCT-13]; and the Special Fund for Basic Scientific Research of Chang'an University [grant number 310821153502 and 310821173501].

References

Bakeer, R.M., et al.: Performance of pile-supported bridge approach slabs. J. Bridge Eng. (2005). https://doi.org/10.1061/(ASCE)1084-0702(2005)10:2(228)

Coelho, B.Z., Hicks, M.A.: Numerical analysis of railway transition zones in soft soil. Proc. Inst. Mech. Eng. [F] J. Rail and Rapid Transit (2015). https://doi.org/10.1177/0954409715605864

Feng, G., et al.: Vertical distance design method utilizing geosynthetic disposing bridge approach roadbed. J. Tongji Univ. (Natural Science) (2003)

Gao, Y., Chen, Y.: Laboratorial studies on dealing with jump at bridge abutment with Netlon geotextile in hignway. J. Changsha Commun. Inst. (1995)

Gao, H., et al.: Vibration Isolation Behavior of EPS Reinforced Highway Embankments. Springer, Berlin Heidelberg (2008). https://doi.org/10.1007/978-3-540-79846-0_77

Ge, Z., et al.: Study actuality summary of backfilling materials behind abutments of highway bridge and culvert. J. Traffic Transp. Eng. (2007)

Han, J., et al.: 2D numerical modeling of a constructed geosynthetic-reinforced embankment over deep mixed columns. Geo-Front. Congr. (2005) https://doi.org/10.1061/40777(156)13

Hu, X., Sun, L.: Stress response analysis of asphalt pavement under measured tire ground pressure of heavy vehicle. J. Tongji Univ. (Natural Science) (2006)

Huang, Y.X.: Pavement Analysis and Design, pp. 190–198. China Communications Press, Beijing (1998)

JTG D50–2017: Specification for Design of Highway Asphalt Pavement. Ministry of transport of PRC, Beijing (2017)

Niu, S.L.: Study on technique of treatment for jump at back of abutment in loess area by the flexible approach slab. Chang'an Univ. (2006)

Robison, J.L., Luna, R.: Deformation analysis of modeling of missouri bridge approach embankments. In: Geotechnical Engineering for Transportation Projects. Asce Geotechnical Special Publication, (2004). https://doi.org/10.1061/40744(154)197

Shi, J., et al.: Measurements and simulation of the dynamic responses of a bridge–embankment transition zone below a heavy haul railway line. Proc. Inst. Mech. Eng. [F] J. Rail and Rapid Transit (2013). https://doi.org/10.1177/0954409712460979

Stark, T.D., et al.: Design and performance of well-performing railway transitions. Transp. Res. Rec. J. Transp. Res. Board (2016). https://doi.org/10.3141/2545-03

Sun, J.: Mechanism analysis of the methods for treating vehicle bumping at bridge approach built on soft foundation and experimental study. Zhe Jiang Univ. (2010)

Thiagarajan, G., et al.: Cost-efficient and innovative design for bridge approach slab. Transp. Res. Rec. J. Transp. Res. Board (2012). https://doi.org/10.3141/2313-11

Wang, H., Al-Qadi, I. L., Stanciulescu, I.: Effect of surface friction on tire–pavement contact stresses during vehicle maneuvering. J. Eng. Mech. **140** (4) (2014). https://doi.org/10.1061/(ASCE)EM.1943-7889.0000691

Yu, Y., et al.: Three-dimensional numerical analysis of geocell flexible approach slab for treating differential settlement at bridge-subgrade transition section. China J. Highw. Transp. (2007)

Zaman, M., et al.: Consolidation settlement of bridge approach foundation. J. Geotech. Eng. (1991). https://doi.org/10.1061/(ASCE)0733-9410(1991)117:2(219)

Zhang, Y.: Research on the pavement behavior and its mixture design for long and steep slope asphalt pavement. Chang'an Univ. (2012)

Zhang, J., Zhang, H.: Additional stress in pavement structure due to asymmetrical settlement of soft subgrade. J. Chang'an Univ. (Natural Science Edition) (2003)

Zhuang, Z., et al.: ABAQUS nonlinear finite element analysis and examples. Science Press, (2005)

Based on Equal Volumetric Parameters Mix Design Method for Foamed Warm Mix Asphalt

Jinjin Shi[1(✉)], Yingbiao Wu[1,2], Qingyi Xiao[3], and Yu Zhang[1]

[1] Cangzhou Municipal Engineering Company Limited, Cangzhou, China
czszjsk@126.com
[2] Hebei Province Road Materials and Technology Engineering Technology Research Center, Cangzhou, China
[3] Hebei University of Technology, Tianjin, China

Abstract. One of the critical issues facing WMA is the lack of a formal mixture design procedure. If warm mix is to replace hot mix in the future, a laboratory mixture design procedure for WMA must be established. The paper developed parametric mix design method for foamed warm mix asphalt based on equal volumetric parameters. Performance comparisons between HMA and WMA were made to study the WMA design method variation. The test results indicated that WMA designed by equal volumetric design method met all specification requirements (ie. stability, flow, bending strain). Thus, the proposed method can be used for routine WMA mixture design. It is concluded that the asphalt foaming process and volumetric design approach did not reduce water stability.

1 Introduction

Warm mix asphalt (WMA) refers to asphalt concrete mixtures that are produced at temperature approximately 30 °C lower than temperature used in hot mix asphalt (HMA). One of the main goals of WMA is to produce mixtures with similar road performance characteristics as HMA but at substantially reduced production temperatures. There are many environmental and health benefits associated with reduced production temperatures [1]. There have been a number of products and processes introduced to the marketplace to produce WMA over the last several years [2–4]. Among these technologies, foamed WMA is the most popular technology.

Wu et al. [reference] played an important role in bringing foamed WMA technologies into practice in China. Numerous demonstration projects have been constructed since 2010 [5, 6]. These projects have demonstrated the feasibility of using foaming warm mix processes in China. In current mechanical plant foaming processes, cold water was injected into a hot binder stream that may be anywhere from 140 °C to 170 °C. The cold water turns to steam when it comes in contact with the hot binder, and the water expanded that creates an increased volume of binder. Foaming works in two ways to promote mixing at lower temperatures. Firstly, it increases the volume of the binder, which makes it easier to coat particles. Secondly, it reduces the overall viscosity of the binder through shear thinning, which makes the mix more workable [7]. But one of the critical issues facing WMA is the lack of a formal mixture design

M. Barman et al. (eds.), *Transportation and Geotechniques: Materials, Sustainability and Climate*, Sustainable Civil Infrastructures,
https://doi.org/10.1007/978-3-319-95768-5_10

procedure [8, 9]. For most WMA projects constructed in China, WMA has been substituted into a mixture designed as HMA without special WMA design procedure, whereas only compaction temperature reduce 20 °C–40 °C compared with that of HMA, which is based on suggestion of supplier. If warm mix is to replace hot mix in the future, a laboratory mixture design procedure for WMA must be established. This paper proposed foamed WMA mixture design and analysis procedures for China conditions and they can be used in wide range of warm mix processes that are currently available or may likely become available in the future.

2 WMA Mixture Design Procedure

Currently, most countries in the world employed Marshall mix design method to determine asphalt mixture proportions [reference]. The essence of Marshall mix design method is volumetric parameters which have a good relationship with road performances. It was widely accepted that volumetric engineering indices had important values to direct asphalt mixture design. So it was excellent choice to design foamed WMA following Marshall volumetric design approach. Using the same origin asphalt, volumetric indices of foamed WMA should be as close to that of HMA as possible. Based on this idea, this study proposed foamed WMA design method which is known as ‘Equal volumetric parameters design method’. In other word, by determination of proper compaction temperature, foamed WMA should have the same volumetric parameter values as HMA.

‘Equal volumetric parameters design method’ has 3 stages. Firstly, according to Marshall test, HMA aggregate gradation, optimum asphalt content and volumetric parameters should be determined. Secondly, volumetric parameters should be obtained from compaction tests of foamed warm mix asphalt at different compaction temperatures and optimum compaction temperature should be determined based on equivalent volumetric parameters. Finally, road properties of warm mix asphalt were tested and must meet standard requirements of HMA with the same asphalt binder. Through comprehensive analyses, the warm mix asphalt design method was characterized with simple principle and practicable procedure.

3 Mixture Design Examples

3.1 Raw Materials

3.1.1 Asphalt

Origin asphalt used in the paper was $70^{\#}$ asphalt (Penetration 60–80) that was made by Binzhou Sino-sea oil refinery. Origin asphalt engineering indices [9] are shown in Table 1.

3.1.2 Aggregate

Limestone aggregate from Mancheng in China was selected for the study, which had high crush value, good angularity, coarse surface texture, low content of flat-elongated particles, high strength, strong adhesion with asphalt and durability.

Table 1. 70# origin asphalt engineering indices

Test item	Test results	Technical requirements
Penetration (25 °C, 5 s, 100 g)/0.1 mm	64.5	60–80
Ductility (15 °C, 5 cm/min)/cm	>100	≥ 100
Soft point/°C	46.4	≥ 46

There were 3 types of coarse aggregates which include 10 mm–15 mm gravel, 5 mm–10 mm gravel, and 3 mm–5 mm gravel. Fine aggregate was 0 mm–5 mm machine made limestone particle. Limestone dust was selected as filler in the study. All engineering indices met 'Specification of construction technology in highway asphalt pavement' [10].

3.2 Mineral Aggregate Gradation

Mineral aggregate gradation design has crucial influences on high temperature stability, moisture damage resistance, and thermal crack resistance. According AC-13 gradation requirement in 'Specification of construction technology in highway asphalt pavement' [11], aggregate particles were composed of relatively low content of nominal maximum size particle and <0.6 mm size particle, and relatively high content of medium size particle, which formed 'S' shape gradation curve, as shown in Table 2. It was found that 2.36mm was the key control size of AC-13 continuous gradation as passing rate on which was below 40%.

3.3 Determination of Optimum Asphalt Content

Based on HMA Marshall design method, optimum asphalt content of AC-13 asphalt was determined. Figure 1 shows the step by step process.

According to aggregate mixture design, asphalt mixtures with 5 different asphalt contents was compacted at prescribed temperature. Then, Marshall volumetric indices were determined as shown in Table 3.

The optimum asphalt content was found to be 4.58%. Volumetric indices at optimum asphalt content (OAC) were summarized in Table 4.

3.4 WMA AC-13 Mixture Design

Some would think that when water is releasing out during foaming process that will cause asphalt to be unstable. However, only approximately 0.1% of water (in asphalt mixture by weight) was added and when no vapor stream in WMA, it allows WMA to perform like HMA. Thus, foaming process can be treated as a compaction aid that allows asphalt mixture to achieve desirable density at lower compaction temperature. It is reasonable to think that WMA design can follow HMA design approach with the same aggregate gradation, OAC, and volumetric requirements.

According to HMA design (aggregate gradation, OAC), aggregate had been weighted, dried and heated. Wirtgen TM WLB10, lab scale foamed asphalt equipment, was utilized in this study. The foaming condition parameters were set at 150 °C for

Table 2. Mineral aggregate gradation design

Aggregate type	Proportion (%)	Passing rate (%)									
		16	13.2	9.5	4.75	2.36	1.18	0.6	0.3	0.15	0.075
10 mm–15 mm	33	100.0	87.1	39.5	0.6	0.6	0.6	0.6	0.6	0.6	0.6
5 mm–10 mm	15	100.0	100.0	95.5	21.5	1.9	1.3	1.3	1.3	1.3	1.3
3 mm–5 mm	12	100.0	100.0	100.0	53.5	18.7	14.3	10.9	9.0	8.0	7.3
Fine aggregate	39	100.0	100.0	100.0	96.8	81.8	48.5	31.6	20.5	17.2	13.7
Filler	1	100.0	100.0	100.0	100.0	100.0	100.0	100.0	98.1	90.3	82.1
Mixture gradation		100.0	95.7	79.4	48.6	35.6	22.0	15.0	10.4	9.0	7.4
Medain value		100.0	95.0	76.5	53.0	37.0	26.5	19.0	13.5	10.0	6.0
Specification requirement	Upper limit	100.0	100.0	85.0	68.0	50.0	38.0	28.0	20.0	15.0	8.0
	Floor limit	100.0	90.0	68.0	38.0	24.0	15.0	10.0	7.0	5.0	4.0

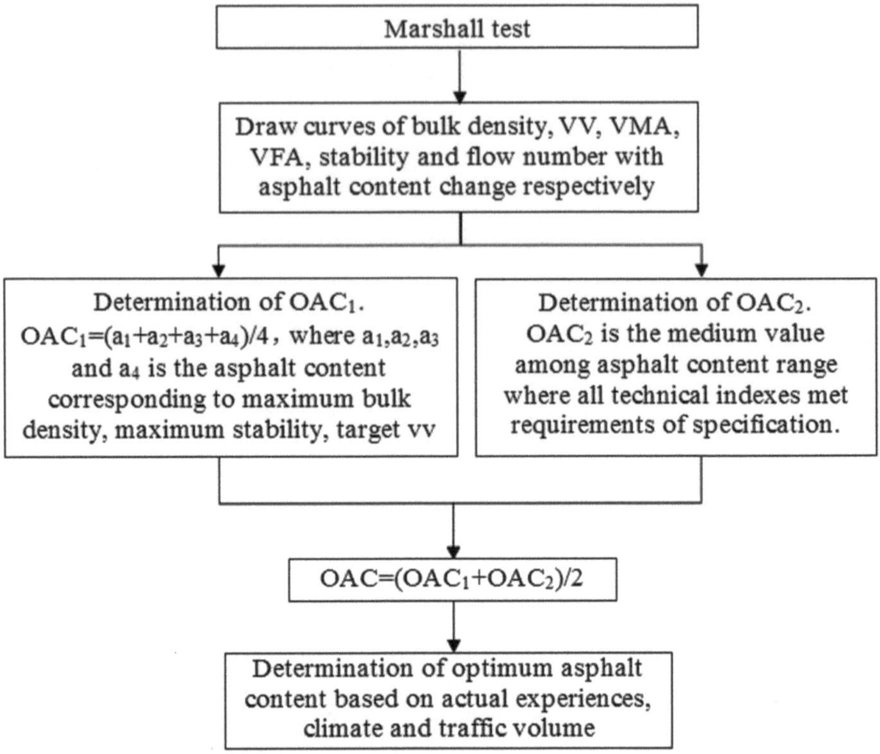

Fig. 1. Determination of optimum asphalt content

Table 3. Volumetric indices of HMA AC-13

Asphalt content (%)	Bulk specific density	Maximum theoretical specific density	VV[a] (%)	VMA[b] (%)	VFA[c] (%)	Stability (kN)	Flow (0.1 mm)
3.38	2.431	2.649	8.222	14.539	43.451	16.66	13.7
3.84	2.426	2.627	7.638	15.125	49.500	15.875	13.1
4.31	2.472	2.606	5.157	13.955	63.045	13.31	14.4
4.76	2.479	2.587	4.172	14.101	70.413	11.57	14.8
5.21	2.481	2.569	3.412	14.429	76.353	10.62	20.7
Specification requirement	–	–	4–6	≥ 14	65–75	≥ 8	15–40

[a]VV is Void of Volume abbreviated.
[b]VMA is Voids in Mineral Aggregate abbreviated.
[c]VFA is Voids Filled with Asphalt abbreviated.

Table 4. Volumetric indices of HMA AC-13 at OAC

Asphalt content (%)	Bulk specific density	VV (%)	VMA (%)	VFA (%)	Stability (kN)	Flow (0.1 mm)
4.58	2.474	4.584	14.110	67.595	11.25	16.7
Specification requirement	–	4–6	≥ 14	65–75	≥ 8	15–40

asphalt binder and 1% foaming water by asphalt binder weight. When aggregate mixture reached setting temperature, foamed asphalt was injected into lab mixer and mixed uniformly with aggregates. Then, WMA mixture was kept in oven for 2 h at specified compaction temperature simulated heating and aging in actual production process (mixing, transporting, and placing). To study the WMA compaction temperature, five different temperatures were utilized (100 °C–140 °C with 10 °C interval). Since workability improvements in WMA, specimens were compacted 75 times each side and they reached almost same volumetric index with HMA mixture. The detailed processing control temperatures for WMA are presented in Table 5.

The volumetric test results of WMA specimens at different compaction temperatures are shown in Table 6.

VV of WMA at 5 different compacted temperatures had been compared with that of HMA, as shown in Fig. 2.

Table 5. WMA processing control temperatures

Processing control temperature	Compaction temperature (°C)				
	100	110	120	130	140
Aggregate heating	120±5	130±5	140±5	150±5	160±5
Foamed asphalt	125±5	125±5	125±5	125±5	125±5
Mixing	110±5	120±5	130±5	140±5	150±5
Curing	100±2	110±2	120±2	130±2	140±2
Compaction	100±2	110±2	120±2	130±2	140±2

Table 6. Test results of WMA specimen at different compaction temperatures

Mixing temperature (°C)	Compaction temperature (°C)	Bulk specific density	VV (%)	VMA (%)	VFA (%)	Stability (kN)	Flow (0.1 mm)
110	100	2.374	7.225	17.583	59.026	6.16	21.9
120	110	2.442	4.562	15.217	70.043	7.59	21.4
130	120	2.441	4.611	15.260	69.835	9.06	23.6
140	130	2.448	4.340	15.020	71.110	9.53	23.0
150	140	2.459	3.895	14.624	73.375	9.89	21.9
Specification requirement		–	4–6	≥ 14	65–75	≥ 8	15–40

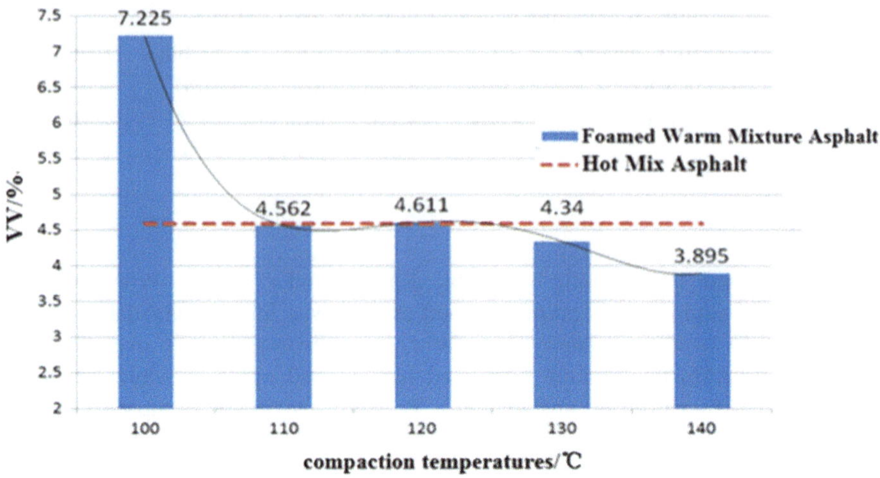

Fig. 2. VV of WMA at different compaction temperatures

Through analyses of aforementioned data, it is concluded that when compaction temperature increases, bulk specific density, VFA and stability increased while VV and VMA decreased. It was observed that there was only minor change on flow number. When compaction temperature was below 110 °C, WMA asphalt had higher viscosity, which reduced mixture workability that led to poorer compaction. Thus, there was less stone-to-stone contact for WMA mixture when temperature was below 110 °C. In addition, when specimen was compacted at 110 °C, it didn't meet specification stability requirement (>8 kN).

When compaction temperature increased from 110 °C to 140 °C, compaction of WMA improved significantly, and VV was close to/below 4%. It was found that 120 ° C was the lowest compaction temperature at which engineering indices met all technical requirements. At the lowest compaction temperature of 120 °C, WMA has better environmental, economic, and social benefits. Thus, 120 °C was selected as compaction temperature in the study.

4 Road Performance Vertification

4.1 High Temperature Stability

High temperature stability of asphalt mixture has been used to evaluate the performance of asphalt concrete pavement. It is very important for engineers to improve pavement performance by analyzing factors that contribute to high temperature stability for asphalt mixtures. Based on Marshall stability and the wheel tracking tests in laboratories, the specimens from HMA and foamed WMA were used to analyze the high temperature performance.

4.1.1 Marshall Stability

Marshall stability and flow results have been used to monitor the plant process of producing asphalt mixture as they are the two main controlling parameters. Marshall stability and flow results of HMA and WMA were shown in Table 7.

Table 7. Marshall test results of HMA and WMA

Mixture type	Compaction temperature (°C)	VV (%)	Stability (kN)	Flow (0.1 mm)
HMA	150	4.584	11.25	16.7
WMA	120	4.611	9.06	23.6
Specification requirement	–	–	≥ 8	15–40

4.1.2 Dynamic Stability

However, Marshall stability and flow are only empirical indices and they may not reflect mixture composite characteristics as asphalt mixtures which met these two requirements still have high probability of rutting [reference]. It is believed that Dynamic stability test can better simulate wheel reciprocating load on the pavement. Thus, dynamic stability test result was employed as high temperature stability index of mixtures. Dynamic stability test results are shown in Table 8.

Table 8. Dynamic stability test results

Mixture type	Compaction temperature (°C)	VV (%)	Dynamic stability (times/mm)
HMA	150	4.565	1369
WMA	120	4.594	1051
Specification requirement	–	–	≥ 1000

Through analysis of Marshall stability and dynamic stability results, high temperature performance of WMA met specification requirements, even though it was a little lower than that of HMA. The difference of HMA and WMA in high temperature stability depends on aging of HMA and WMA production. Because HMA was produced at higher temperature, degree of aging in HMA was much higher than that in WMA.

4.2 Low Temperature Anti-cracking Resistance

Thermal cracking at lower temperature was one of the main distresses in asphalt pavement. During winter time, the temperature of pavement structure in northern China drops significantly. When temperature decreases, asphalt pavement thermal stress will relax because of asphalt binder viscoelastic characteristics. If surrounding temperature decreases rapidly, thermal stress in the pavement will increase without time for stress relaxing. When thermal stress equal to or greater than tensile strength of asphalt mixture, low temperature thermal crack distress will occur on the road, which reduces pavement service level and structure durability.

In China, bending beam test was the standard (JTG F40) test to evaluate cracking resistance of asphalt mixture at low temperature. It is a 3-point bending beam test that includes 3 technical indices: (1) tensile strength, (2) failure bending strain, and (3) stiffness (modulus). Failure bending strain reflects deformability of asphalt mixture at low temperature which has good relationship with thermal cracking resistance and thus it is used as a control index in asphalt mixture design. Bending beam test results are shown in Table 9.

Table 9. Bending beam test results

Mixture type	Compaction temperature (°C)	VV (%)	Tensile strength (MPa)	Failure bending strain (με)	Stiffness modulus (MPa)
HMA	150	4.329	5.69	2074	2745.52
WMA	120	4.416	5.59	2249	2708.85
Specification requirement	–	–	–	>2000	—

These results shows failure bending strain of WMA was more than that of HMA, which mean WMA had better thermal crack resistance. This phenomenon is because of WMA was produced and compacted at lower temperature.

4.3 Water Stability

Like rutting and thermal cracking, moisture damage is another critical parameter that reduces pavement service level and pavement life. Moisture typically reduces stiffness of the binder and/or mastic through moisture diffusion that degrades the adhesive bonding between the binder/mastic and aggregate particles. A loss of HMA internal strength will result in premature distresses such as rutting, raveling, and fatigue cracking. At present, immersion Marshall test and Lottoman test (AASTHO T283) were employed to evaluate water stability of WMA and HMA. Results are shown in Tables 10 and 11.

Table 10. Immersion Marshall test results

Mixture type	Compaction temperature (°C)	VV (%)	Marshall stability (kN)	Marshall stability after 48 immersion (kN)	Retained Marshall ratio (%)
HMA	150	4.584	11.25	10.94	97.22
WMA	120	4.611	9.06	8.83	97.46
Specification requirement	–				≥ 80

Table 11. Lottoman test results

Mixture type	Compaction temperature (°C)	VV (%)	Splitting strength (MPa)	Splitting strength after freeze-thaw (MPa)	Retained strength ratio (%)
HMA	150	4.584	1.50	1.34	89.33
WMA	120	4.611	1.16	1.01	87.35
Specification requirement	–				≥75

In view of Tables 10 and 11, it was found that the water stability of WMA met all specification requirements. Also, the results from WMA were very similar to that of HMA. The results indicated that the asphalt foaming process and volumetric design method did not reduce water stability.

5 Conclusions

Based on traditional asphalt mixture design method, this paper proposed equal volumetric parameter design method for foamed WMA. In addition, comparisons between HMA and WMA were made to study performance variation. Based on the analyses, observations and conclusions are given as follows:

(1) The essence of the design method was that WMA compaction temperature should be determined based on the principle that volumetric indices of WMA was equal or close to those indices of HMA.
(2) The test results indicated that WMA designed by equal volumetric design method met all specification requirements (ie. stability, flow, bending strain). Thus, the proposed method can be used for routine WMA mixture design.
(3) The results revealed that the asphalt foaming process and volumetric design approach did not reduce water stability.

References

1. Prowell, B.D., Hurley, G.C.: Warm-Mix Asphalt Best Practices. Quality Improvement Series 125. National Asphalt Pavement Association, Lanham, MD (2007)
2. Hanz, A., Faheem, A., Mahmoud, E., Bahia, H.: Measuring Effects of Warm-Mix Additives Using a Newly Developed Asphalt Binder Lubricity Test for DSR. Transportation Research Board 89th Annual Meeting Compendium of Papers DVD, Transportation Research Board of the National Academies, Washington, DC (2010)
3. Csanyi, L.H.: Foamed Asphalt in Bituminous Paving Mixes. Bulletin 160. Highway Research Board, National Research Council, Washington, DC, pp. 108–122 (1957)
4. Brashears, D.: Warm Mix Hot Again. Innovations, vol. 1, no. 2, pp. 14–17. Gencor Industries, Inc., High Velocity Communications, Inc., Waukesha, WI (2007)

5. Yingbiao, W., Yanfang, G., Jinjin, S., Baofeng, Z.: Application of Warm Mix Technology Based on Foamed Asphalt in Road Construction, vol. 4, no. 29, pp. 17–21. Municipal Engineering Technology, China (2011)
6. Yang, Y., Wu, Y., Qiao, J.: Experimental Study on Aspha-min Warm-mix Binder Performance, vol. 3, no. 34. Chinese and Foreign Road, China, pp. 291–294 (2014)
7. Fort, J.-P., G. Graham, J. Farnham, S. Sinn, Rowe, G.M.: I-55/I-57 Warm Mix Project in Missouri Contractor's WMA Experience. Proceedings. 2nd Int'l Conf on Warm Mix Asphalt. National Asphalt Pavement Association, Lanham, MD. www.asphaltpavement.org. Accessed 22 Oct 2011 (2011)
8. Advanced Asphalt Technologies, LLC.: Draft Final Report on National Cooperative Highway Research Program Project 09-33: A Mix Design Manual for Hot Mix Asphalt. (2009)
9. Highway Research Institute of the Transportation Department.: Standard Test Methods of Bitumen and Bituminous Mixtures for Highway Engineering (JTG E20-2011), pp. 186–356. China communications Press, Beijing (2011)
10. Highway Research Institute of the Transportation Department.: Technical Specifications for Construction of Highway Asphalt Pavements (JTG F40-2004), pp. 23–39. China communications Press, Beijing (2004)
11. Highway Research Institute of the Transportation Department.: Test Methods of Aggregate for Highway Engineering (JTG E42-2005), pp. 8–133. China communications Press, Beijing (2005)

Use of Falling Weight Deflectometer for Airport Pavements

Greg White(✉)

Airport Pavement Research Program, University of the Sunshine Coast, Sippy Downs, Queensland, Australia
gwhite2@usc.edu.au

Abstract. The Falling Weight Deflectometer (FWD) provides a rapid and cost-effective means of measuring the response of pavements to a dynamic load. Its broad application has allowed the development of significant guidance regarding empirical interpretation of FWD results. Software has also been developed allowing the estimation of layer modulus, including the subgrade, and in some cases the calculation of a theoretical strength rating for aircraft pavement structures. However, many practitioners question the reasonableness of the moduli calculated and the reliably of the resulting airport pavement strength rating. This paper reviews the use of FWD in airport pavement strength evaluations and assesses its appropriateness for particular situations and tasks. Example projects are considered and data analysed to determine the benefits and limitations. Further work is recommended to better understand the impact of variability in FWD results on inferred layer modulus and airport pavement strength rating.

1 Introduction

With airports getting busier and upgrade funding becoming tighter, Non-Destructive Testing (NDT) which is rapid to perform and relatively economical to procure, has gained popularity. Particularly for busy airport runways that can not readily be closed for days of traditional intrusive pavement investigation (White 2017). The Falling Weight Deflectometer (FWD) is arguably the most common NDT device used for the structural evaluation of existing airport pavement infrastructure (Celaya and Nazarian 2014).

FWDs have been in use since the 1980s. These generally trailer-mounted NDT devices apply a dynamic load while the trailer is stationary. The load can be varied and deflections are estimated from surface strain accelerations measured by geophones placed at various distances from the load application (TRB 2008). This results in a deflection profile or bowl that is a function of the composition and condition of the pavement (Vuong 1989).

This paper reviews the use of FWD in airport pavement strength evaluations and assesses its appropriateness for particular situations and tasks. Example projects are presented and data analysed to determine the benefits and limitations. Further work is recommended to better understand the impact of variability in FWD responses on inferred layer modulus and airport pavement strength rating.

M. Barman et al. (eds.), *Transportation and Geotechniques: Materials, Sustainability and Climate*, Sustainable Civil Infrastructures,
https://doi.org/10.1007/978-3-319-95768-5_11

2 Background

2.1 Airport Pavement Strength Rating System

Airport pavements are usually either flexible or rigid structures, although composite pavements are also used in some circumstances. Rigid airfield pavements are often plain jointed concrete on a stabilised sub-base. The joints are required to transfer loads, either by aggregate interlock across sawn contraction joints or by dowels across formed construction joints. Flexible pavements usually comprise significant thickness of crushed rock base and sub-base with a thin (50–100 mm) or thick (100–250 mm) asphalt surface, although some regional airports use a bituminous sprayed seal wearing surface.

Both rigid and flexible airport pavements are subject to an international strength rating system known as ACN-PCN (ICAO 1983). The system is similar to road axle load limits but reflects the higher level of deviation between small aircraft and large aircraft, meaning every airport has its own Pavement Classification Number (PCN), which represents the upper limit of Aircraft Classification number (ACN) allowed for unrestricted operations.

Traditionally an airport's PCN was determined by reverse engineering the existing pavement structure, including layer thicknesses and materials, as well as the subgrade CBR, to determine the acceptability of a particular aircraft operating at a particular mass. The PCN was generally set to the highest ACN of all the aircraft considered to be acceptable for the particular pavement. This approach remains the normal method for PCN assignment in many countries (FAA 2011). However, as detailed below, software is now available to automatically calculate a PCN, primarily from FWD deflection data.

2.2 Falling Weight Deflectometer

There are at least four manufacturers of FWD devices, although Dynatest is the most commonly encountered brand. All FWDs apply a vertically dynamic load onto a 300–450 mm diameter load plate. The duration of the dynamic loading is generally 20–65 ms (Ameri et al. 2009). The load generally ranges from 7 to 150 kN for the standard FWD, although a light version (LWD) applies 1–15 kN and various heavy versions (HWD) apply up to 250 kN (FAA 2011). Deflections are estimated from geophone measurements of surface acceleration, that can be spaced as required, usually anywhere from 0 to 2400 mm from the centre of the loading plate. Standard distances for geophone locations are 0, 300 600, 900, 1200, 1500 and 1800 mm from the centre of the load plate. However, different countries adopt different spacings. For example, Australia specifies a minimum of seven geophones, with three located 0, 200 and 900 mm from the load plate centre with the location of the other geophones optional (Austroads 2011). This approach reflects the focus on the deflections at 0, 200 and 900 mm for the characterisation of deflection bowl, using the generic term 'D_n' to represent the deflection at the geophone located n mm from the centre of the load pate.

Deflection bowls are generally characterised by three parameters (Austroads 2011)

- Maximum deflection (D_0). A general indicator of pavement stiffness and response.
- Curvature (D_0–D_{200}). An indicator of the upper base course and surface layer stiffness.
- D_{900}. An indicator of the subgrade support condition.

The above indicators are only general in nature. For example, a very thick pavement's subgrade is less likely to be adequately characterised by D_{900} because the measured deflection at 900 mm from the load plate centre will be less affected by the subgrade when the subgrade is located much deeper than typical.

2.3 Modulus Back Calculation

A number of softwares are now available for the back calculation of layer modulus values from deflection bowl data. Example softwares include Modulus, Elmod, Evercalc (Ameri et al. 2009), DAPS (Bandara et al. 2002), Elsedef, Illi-Back (Kang 1998) and EfromD (Vuong 1989). Generally, the software requires pavement layer thicknesses to be nominated as well as modulus seed values. Although attractive to the designer, absolute back calculated modulus values should be used with caution (Gendreau and Soriano 1998) due to the high degree of variance between reported and observed pavement performance (Zaniewski 1991). Importantly, the nominated layer thicknesses can significantly affect the resulting layer modulus values, as detailed later.

Due to the prevalence of the Dynatest FWD, Dynatest's ELMOD (Dynatest 2017) is arguably the most common back analysis software used by practitioners. The latest version, ELMOD 6, also includes PCN determination directly from deflection bowl data.

Following layer modulus estimation, the software uses nominated aircraft type and frequency to determine the most damaging aircraft for each nominated pavement layer. The ACN of the most damaging aircraft at the 'just acceptable' aircraft mass is reported as the allowable pavement PCN. The allowable mass may be higher than the published maximum mass of the aircraft in order to 'just fail' the pavement by the end of the design life. The pavement damage caused by each aircraft is based on an allowable stress or strain levels for each pavement layer. This approach does not take into account superposition of the impact of multiple aircraft types, relies upon the allowable stress/strain determined for various pavement materials and relies upon the accuracy of the pavement layer thicknesses.

In ELMOD 6, every test location has a PCN value reported. The PCN values vary and a protocol is required for determining a characteristic PCN from the hundreds of test points.

3 Example applications

3.1 Homogenous Sectioning

Arguably, the most common application for FWD surveys in airport pavements is the identification of sections of pavement of homogenous response to load. Sections of homogenous response inform intrusive testing locations for increased efficiency. Rather than randomly locating intrusive test points, the intrusive testing is planned to reflect the designated sections of homogenous response to load. The approach increases efficiency for pavements that are operationally difficult or expensive to close for intrusive testing.

Maximum deflection (D_0) is often used as the basis of homogenous sectioning. However, some examples have also used CV and D_{900} in combination with D_0. An

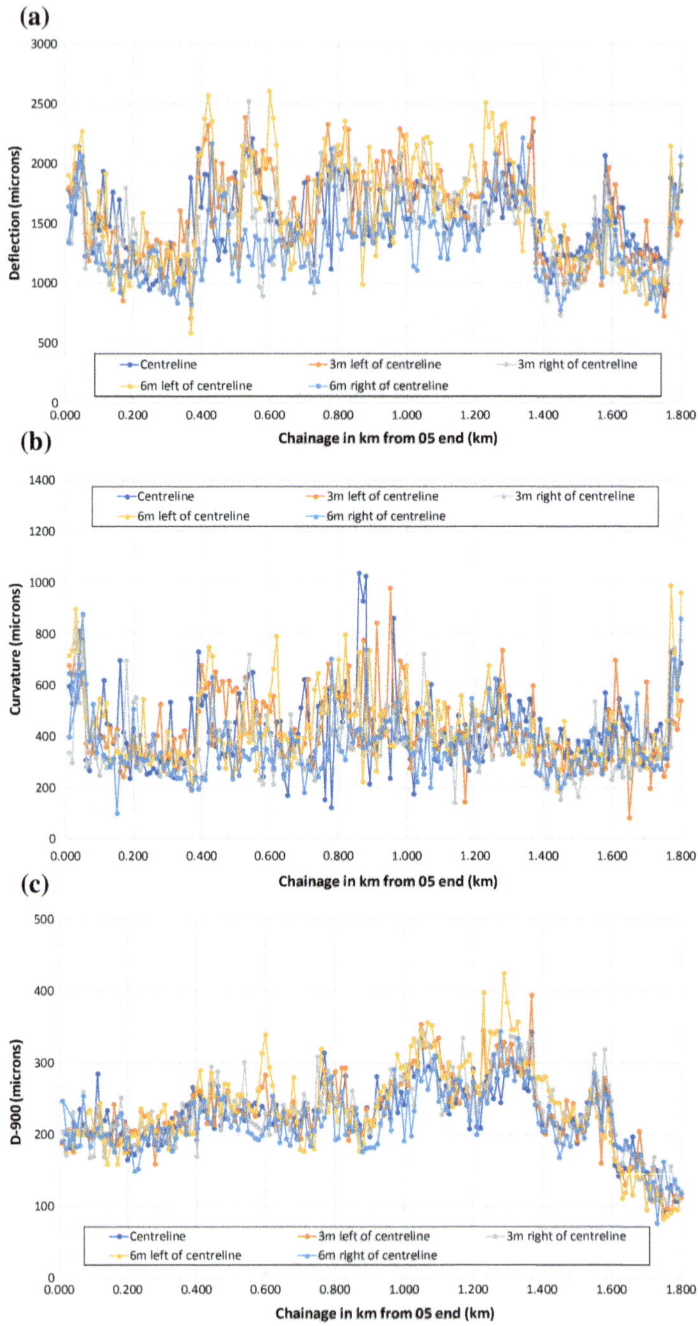

Fig. 1. **a** Example pavement response—D_0. **b** Example pavement response—CV. **c** Example pavement response—D_{900}

example is shown in Fig. 1 where D_0 analysis (Fig. 1a) indicates clear changes in response at approximately CH400 and CH1400. The 1000 m in between is very consistent. Figure 1b (CV) indicates that the upper base course is generally consistent along the full length of the runway, while the subgrade response (D_{900}) indicates the changes in D_0 most likely reflect changes in the subgrade condition (Fig. 1c). Intrusive testing subsequently verified this finding, encountering a change in subgrade material type at CH1400.

Identifying pavement areas of homogenous response is often performed visually, however, statistical tools are also available. Cumulative sum analysis was developed in the 1950s for statistical quality control of sequential processes developed. The cumulative differences method is an adaptation recommended by the US Highways Associations (AASHTO 1993) designed for identifying trends in various pavement responses (Thomas 2004). Conveniently, cumulative differences graphs start and end at 0, and changes in gradient indicate changes in pavement response trend, as shown in the example at Fig. 2 for D_0 values along a runway centreline and at three different offsets either side of the centreline. The cumulative difference chart indicates changes in overall pavement response at chainages 700 m and 1700 m, resulting in three homogenous sections. A similar analysis was also undertaken on CV and D_{900} values. One disadvantage of the cumulative differences approach is a reduction in intuitive interpretation of the results, reflecting the 'relative to the average' nature of the values.

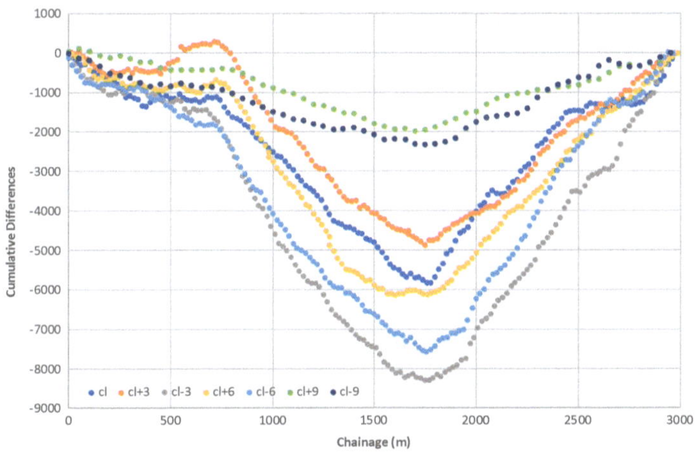

Fig. 2. Example cumulative difference for D_0

3.2 Pavement Layer Modulus Estimation

As discussed above, various softwares are available for estimating layer modulus from deflection bowl data. ELMOD 6 is currently the most common, reflecting the prevalence of the Dynatest FWD equipment. Like other softwares, ELMOD 6 requires the layer thicknesses to be nominated, as well as seed values of modulus.

Depending on the software used, two main approaches are possible. The first returns an estimated modulus value for each recorded test location. A representative value is then determined by selecting a percentile-based representative value of modulus for each layer, with 10 and 15 percentile values common. The second approach requires all the deflection bowl data to be analysed, a representative deflection bowl adopted and the representative deflective bowl converted to estimated modulus values for the various layers. This approach requires more interpretation of the deflection data. Consequently, practitioners not familiar with FWD analysis often prefer the percentile-value approach, which requires less judgment.

The data presented in Fig. 1 was analysis by ELMOD 6, assuming a four-layer system (40 mm asphalt, 160 mm base, 200 mm sub-base and subgrade). Table 1 summarises the data for the three homogenous sections. The variability is concerningly high, with coefficients of variation (CoVs) ranging from 30% to 160%. Similarly, the modulus of the surface layer varies significantly, despite all surface layers at the airport being constructed and surfaced as a single project. The estimated modulus values range from 72 MPa to 48,900 MPa and the interquartile range (difference between 75 and 25 percentile values) divided by the median is 0.44 to 0.59 across the three homogenous sections, which is high.

Table 1. Summary of estimated modulus values

Statistic	40 mm asphalt	160 mm base	200 mm sub-base
Homogenous Section CH 0-370			
Minimum	3,971	62	48
25 percentile	13,501	118	186
Average	16,769	158	517
Median	16,709	144	298
75 percentile	20,925	232	441
Maximum	30,916	478	6,217
Standard deviation	5,050	67	679
CV	30.1%	42.5%	131.5%
Number of results	185		
Homogenous Section CH 380-1370			
Minimum	72	53	27
25 percentile	9,925	137	79
Average	13,290	180	160
Median	12,551	168	112
75 percentile	15,752	209	165
Maximum	34,319	764	3,943
Standard deviation	5,123	71	257
CV	38.5%	39.1%	160.4%
Number of results	500		

(*continued*)

Table 1. (*continued*)

Statistic	40 mm asphalt	160 mm base	200 mm sub-base
Homogenous Section CH 1380-1820			
Minimum	2,329	49	47
25 percentile	11,153	124	159
Average	16,293	184	351
Median	15,900	173	257
75 percentile	20,488	231	441
Maximum	48,931	428	2,931
Standard deviation	6,936	82	312
CV	42.6%	44.6%	88.8%
Number of results	225		

3.3 Subgrade CBR Estimation

ELMOD 6 also estimates modulus values for the subgrade layer, which can be used to estimate a subgrade CBR value. There are many equations for the estimation of subgrade CBR from modulus, and vice-versa, with CBR = Modulus/10 commonly adopted for airport pavement design purposes. However, ELMOD 6 uses a power-law conversion as shown in Fig. 3, approximated as CBR = $0.0113 \times \text{Modulus}^{1.5625}$. This significantly increases the subgrade CBR over the typical airport pavement design CBR range of 3–15, by up to 100%.

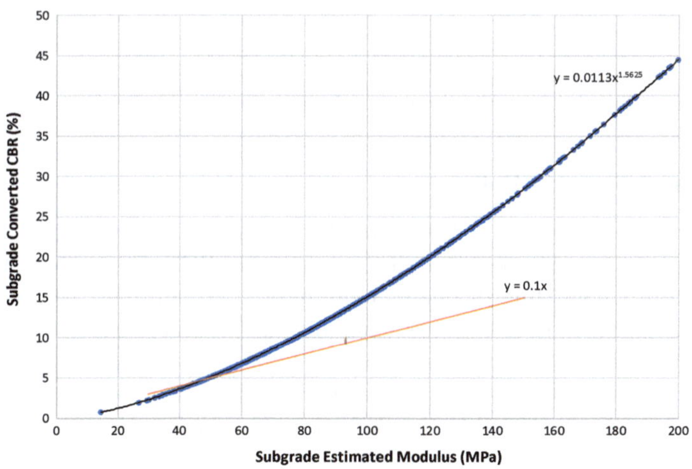

Fig. 3. Inferred ELMOD 6 subgrade CBR from Modulus

Like the modulus values, the CBR values are concerningly variable with even the most homogenous section from Fig. 1 (CH 380-1370) having a CoV of 67% for estimated CBR with values ranging from 1 to 61. This range of CBR values is highly unlikely within a generally uniform site and a runway of generally sound performance.

As another example, the estimated 10 percentile CBR values, by homogenous section and centreline offset, for the airport represented in Fig. 2, is shown in Fig. 4. Although some variation along the runway length is accepted, it is unlikely that the subgrade CBR actually varies from above 20 at offsets 3 m, 6 m and 9 m either side of the runway, to just 6 along the centreline. Further, the performance of the airfield did not reflect this variation in performance under aircraft traffic, adding weight to the belief that the data is not representative of the true subgrade condition. It is hypothesized that testing on the crown may have affected the results by preventing uniform contact between the FWD loading plate and the pavement surface. However, further work is required to better understand the potential source of these illogical results.

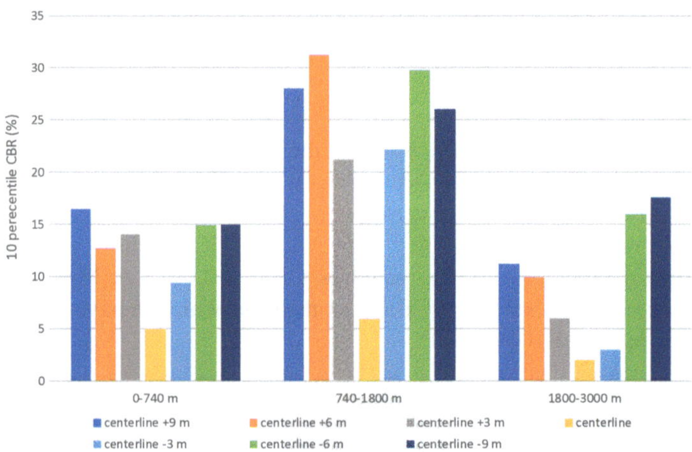

Fig. 4. Example estimated subgrade CBR by FWD offset

3.4 Pavement Strength Rating

Once the subgrade CBR and the layer modulus values are determined, ELMOD 6 can also determine a pavement strength rating, expressed as PCN. To complete the process, a traffic spectrum, consisting of one or more design aircraft, must be nominated. As summarised in Table 2, the PCN values are almost as variable as the modulus values. The calculated PCN values range from 2 to 27 and the 25 and 75 percentile values are 4 to 10. These are significant deviations in PCN considering the damaging effect of aircraft is exponentially related to the PCN (Fig. 5). For the runway in question, the published PCN is 14, near the maximum value calculated by ELMOD 6, which is actually the 93 percentile value. This compares to common practice which would recommend the 15 percentile value, in this case 4. Therefore, given the generally good performance of the runway, adopting the ELMOD 6 recommended PCN would significantly and unjustifiably limit to use of the runway.

Table 2. Summary of calculated PCN values

Statistic	CH 0-370	CH 380-1370	CH 1380-1820
Minimum	4	2	3
25 percentile	8	4	6
Average	11	6	8
Median	10	5	8
75 percentile	10	7	10
Maximum	27	16	20
Standard deviation	4	2	3
CV	38.3%	44.3%	42.0%
Number of results	185	500	225

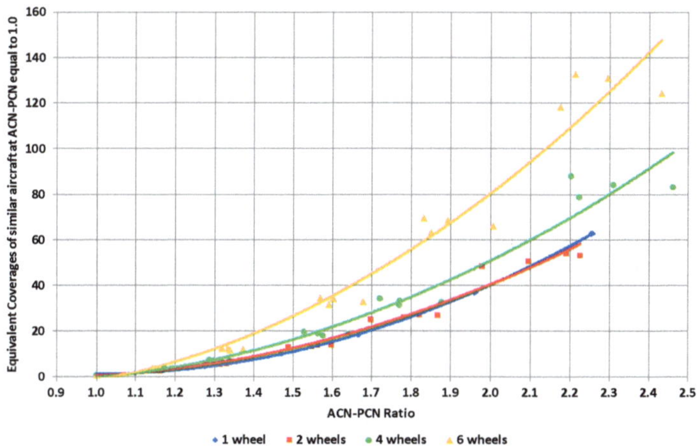

Fig. 5. Typical aircraft overload damage as a function of ACN ratio

ELMOD 6 required a nominated number of layers and a thickness for each. Consequently, for a given aircraft, the estimated layer modulus values and subgrade CBR are the only other factors that will impact the PCN value calculated by ELMOD 6. Because aircraft pavement design is highly sensitive to subgrade CBR (White 2005) the PCN values are expected to be highly correlated to the estimated subgrade CBR values. However, this is not the case, as shown in Fig. 6. This can only reflect the unrealistically high influence of the estimated asphalt, base and sub-base modulus values of the resulting ELMOD 6 PCN, because these are the only other factors affecting the calculated PCN values.

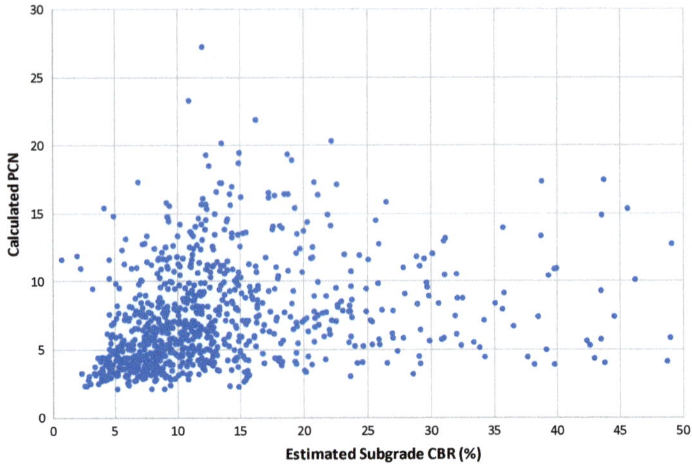

Fig. 6. ELMOD 6 recommended PCN as a function of calculated subgrade CBR

3.5 Concrete Joint Load Transfer

Rigid airfield pavement thickness design determines a centre-slab thickness of concrete. Stress concentrations at slab joints require an additional 25% of the centre-slab thickness to withstand the increased magnitude of stress. However, load transfer from the loaded slab to the adjacent slab is relied upon to avoid the increased concrete thickness. In practice, longitudinal construction joints are usually dowelled to affect load transfer, while transverse sawn contraction joints rely on aggregate interlock. In older pavements, load transfer efficiency is often assessed by FWD.

FWD analysis of load transfer efficiency is usually expressed by the ratio of FWD deflection measured 300 mm either side of the FWD load plate, with the FWD orientated so that the joint is located between the load plate and the D_{300} geophone. In practice, the transverse and longitudinal joints, the corner of the slab and the centre of the slab are all assessed, as illustrated by Fig. 7.

Example rigid airfield pavement joint load transfer survey results are summarised in Table 3 and illustrated as box-and-whisker plots in Fig. 8. The near-perfect symmetry of responses measured 300 mm either side of the load plate is demonstrated by the average ratio of 98% and low standard deviation of 3%. The longitudinal joints exhibit excellent load transfer, with an average deflection ratio of 98%, while the transverse joints are significantly less efficient with an average deflection ratio of 45% and a standard deviation of 28%. This represents the less reliable load transfer by aggregate interlock across old sawn joints, compared to that of dowels. The slab corner results are similar to those for the transverse joints.

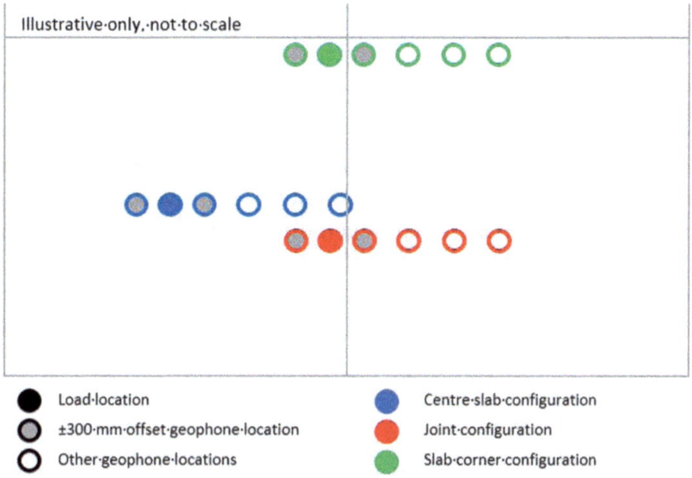

Fig. 7. Typical joint transfer efficient assessment configurations

Table 3. Summary of joint load transfer efficiency

Statistic	Centre of Slab	Longitudinal Joints	Transverse Joints	Corner of Slab
Minimum	83%	90%	3%	3%
25 percentile	98%	97%	21%	30%
Average	98%	99%	45%	55%
Median	99%	99%	39%	50%
75 percentile	100%	102%	58%	81%
Maximum	107%	109%	113%	136%
Standard deviation	3.1%	3.3%	28.4%	33.1%
Number of results	146	74	74	74

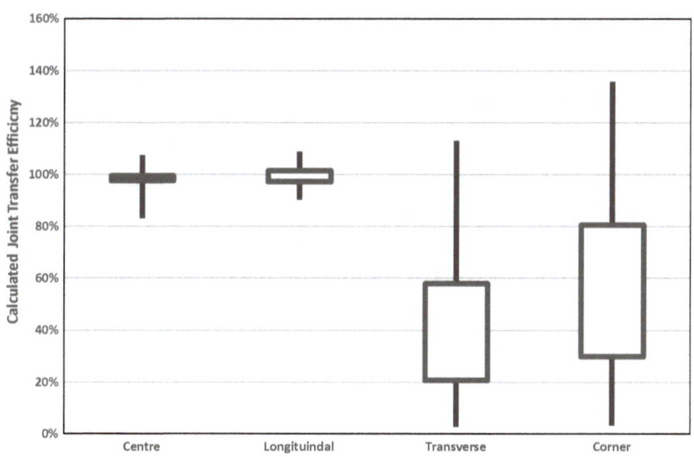

Fig. 8. Example calculated joint transfer efficiency summary

4 Limitations

4.1 Repeatability of Response

Repeatability of the response is believed to account for some of the variability in pavement response, as detailed below. The repeatability of a single FWD machine and the reproducibility across a fleet of six FWDs in the USA indicated the importance of standardised calibration to improve reproducibility (Rocha et al. 2004). The study also identified reduced repeatability as the distance of the geophone from the load plate increased. A simple comparison of repeated test configuration and location within the data presented in Fig. 1 supports these findings, with good agreement between the first and repeated drop, with a linear correlation of R = 96%, although some significant outliers were identified Fig. 9. Further work is recommended to understand the typical reproducibility of FWD results.

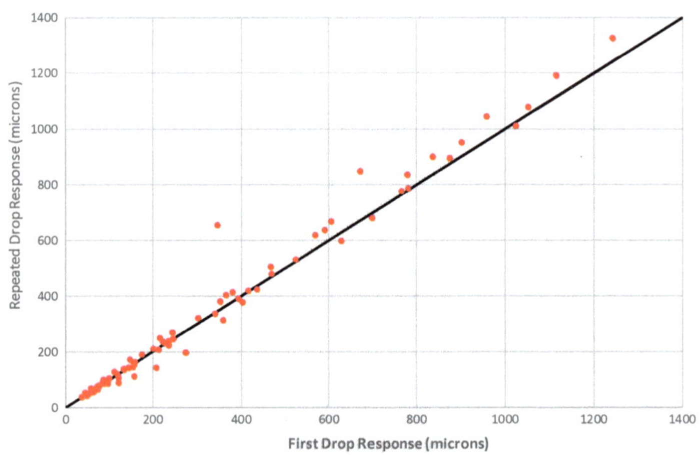

Fig. 9. Example comparison of repeated FWD response

4.2 Variability of Pavement Response

Generally homogenous sections of pavement, of nominally consistent construction, indicate significant variability in pavement responses. For example, the homogenous section of pavement represented in Fig. 1 (CH 380-1370) exhibits high variability in the measured D_0, curvature and D_{900}, with CoV values of 19%, 31% and 17%, respectively (Table 4). Further work is required to understand if this level of variability is typical and its implications for airport pavement evaluation.

Table 4. Summary of measured deflection values

Statistic	D_0	CV	D_{900}
Minimum	894	170	122
25 percentile	1464	226	344
Average	1690	255	431
Median	1691	249	411
75 percentile	1896	284	498
Maximum	2607	425	1038
Standard Deviation	500	500	500
CV	317	42	132

4.3 Reliance on Input Assumptions

As detailed above, the existing pavement layer thicknesses must be input into ELMOD 6 for modulus and PCN calculation. The assumed layer thickness has a significant effect on the estimated modulus value. For example, a light aircraft runway was surveyed by FWD but historical records could not determine whether the existing base course was 150 mm or 250 mm thick. Layer modulus values were determined by ELMOD 6 using both thicknesses (Fig. 10). When the greater base course thickness was assumed, both the base course and subgrade modulus reduced by an average of approximately 30% (Table 5). However, the reduction in estimated modulus was not consistent. Further work is required to understand if this level of variability is typical for airport pavement FWD surveys. Further, the sensitivity of pavement life to pavement thickness (White 2005) results in the layer thickness assumption having a significant effect on the ELMOD 6 calculated PCN values. For the previous example, the increased base course thickness assumption resulted in an average PCN increase of 74%, despite the decrease in base and subgrade modulus (Table 5). Further work is required to better understand the impact of input assumptions on the estimated modulus and PCN values.

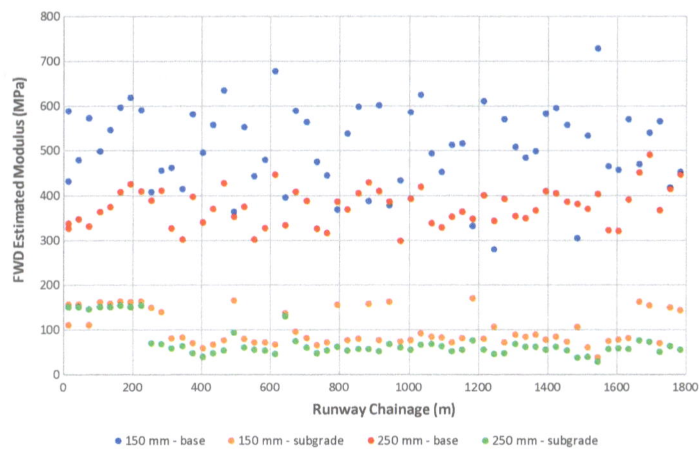

Fig. 10. Example effect of base thickness on base and subgrade modulus

Table 5. Layer thickness impact on estimated modulus and PCN

Statistic	Base Modulus (MPa)	Subgrade Modulus (MPa)	PCN
Minimum	0.55	0.36	1.38
25 percentile	0.68	0.66	1.66
Average	0.76	0.71	1.77
Median	0.69	0.73	1.74
75 percentile	0.73	0.76	1.83
Maximum	1.25	1.35	2.45
Standard Deviation	60	60	60
CV	0.16	0.19	0.23

5 Conclusions

FWD devices provide a rapid and cost-effective means of assessing airport pavement response to load. Although reported to be reasonably repeatable, reproducibility between devices is less certain. The significant response variability observed in generally homogenous sections of pavement is concerning and further work is recommended to understand whether the variability of the presented data is typical. Additional work is also recommended to determine the impact of FWD response variability and input assumptions on back-analysed layer moduli and airport pavement PCN values. Understanding these issues will further increase the usefulness of FWD surveys in airport pavement assessment and management.

Acknowledgments. The provision of FWD data and advice regarding testing protocols provided by Jim Johnson-Clarke or ARRB (Melbourne, Australia) is gratefully appreciated.

References

AASHTO: Guide for Design of Pavement Structure, Appendix J: Analysis Unit Delineation by Cumulative Differences. American Association of State Highway and Transportation Official, Washington, District of Columbia, USA (1993)

Ameri, M., Yavari, N., Scullion, T.: Comparison of static and dynamic backcalculation of flexible pavement layer moduli, using four software programs. Asian Journal of Applied Sciences **2**(3), 197–210 (2009)

Austroads.: Pavement Deflection Measurement with a Falling Weight Deflectometer. Austroads Test Method AG:AM/T006, 31 Mar 2011

Bandara, N., Rowe, G.M., Sharrock, M.J., Nickerson, C.R.: Seasonal variation of subgrade modulus in different subgrade soils for pavement rehabilitation for non freeze-thaw cycles. In: Applications of Advanced Technology in Transportation, pp. 473–480 (2002)

Celaya, M., Nazarian, S.: 'Field evaluation of NDT devices for delamination detection of HMA airport pavements. In: FAA Worldwide Airport Technology Transfer Conference, Galloway, New Jersey, USA, 5–7 Aug 2014

Dynatest 2017, *ELMOD 6 Quick Start Manual*, Dynatest International

FAA.: Use of Nondestructive Testing in Pavement Evaluations, Advisory Circular 150/5370-11B, Federal Aviation Administration, Washington, District of Columbia, USA, 30 Sept 2011
Gendreau, M., Soriano, P.: Airport pavement management systems: an appraisal of existing methodologies. Transp. Res. **32**(3), 197–214 (1998)
ICAO: Aerodrome Design Manager: Part 3, ICAO 9157, 2nd edn. International Civil Aviation Organization, Montreal, Quebec, Canada, January (1983)
Kang, Y.V.: Multifrequency back-calculation of pavement layer moduli. J. Transp. Eng. **124**(1), 73–81 (1998)
Rocha, S., Tandon, V., Nazarian, S.: Falling weight deflectometer fleet. Road Mat. Pavement Des. **5**(2), 215–238 (2004)
Thomas, F.: Generating homogenous road sections based on surface measurements: available methods. In: 2nd European Pavement and Asset Management Conference, Berlin, Germany, 21–23 Mar 2004
TRB: Falling Weight Deflectometer Usage, NCHRP Synthesis 381. Transportation Research Board, Washington, District of Columbia, USA (2008)
Vuong, B.: A new linear elastic back-calculation model for back-calculating layer moduli at fixed Poisson's ratio. Aust. Road Res. **19**(1), 17–28 (1989)
White, G.: A sensitivity analysis of APSDS, an Australian mechanistic design tool for flexible aircraft pavement thickness determination. In: First European Aircraft Pavement Workshop, Amsterdam, Netherlands, 11–12 May 2005
White, G.: Expedient runway upgrade technologies. In: 10th International Conference on the Bearing Capacity of Roads, Railways and Airfields, Athens, Greece, 28–30 June 2017
Zaniewski, J.: Unified Methodology for Airport Pavement Analysis and Design, Technical Report DOT/FAA/RD-91/15, I. Federal Aviation Administration, Washington, District of Columbia, USA, June (1991)

Studies on Performance of Latex Based Coatings for Concrete Surfaces

M. Sarumathi[1], S. N. Ramaswamy[1(✉)], and R. Selvaraj[2]

[1] Department of Civil Engineering, Kalasalingam University, Krishnankoil, Tamil Nadu, India
{sarumathi.m,s.n.ramaswamy}@klu.ac.in

[2] Material Protection and Corrosion Division, CSIR-CECRI, Karaikudi, Tamil Nadu, India
selvarajcecri@gmail.com

Abstract. In this paper, we have demonstrated the effectiveness of latex based coatings applied over concrete surfaces by conducting water absorption test, abrasion test, chloride ion penetration test and heat resistance test. Fifteen types of water based coatings are formulated using water proofing latex as binder along with fly ash, Ordinary Portland Cement, feldspar, high alumina clay, kaolin, Micro silica, black clay, Zinc phosphate, alumina, Titanium dioxide, silica fume and clay as functional filler materials. Water permeability is a major issue in concrete structures. Thus, water permeability was assessed by immersing the specimen of size 40 mm diameter and 100 mm long in water for 48 h. Similarly, abrasion resistance was tested by scribing the coated surface of 100 × 100 × 10 mm tile specimen using E4 320 emery sheet. The most important chloride ion penetration was investigated according to ASTM C1202 by immersing disc specimen of size 80 mm diameter and 40 mm long in 3% Sodium Chloride solution in one end and 0.1 M Sodium Hydroxide solution in the other end. Finally, the heat resistance was determined by placing the 230 × 230 × 20 mm coated tile in a top loading type muffle furnace, by maintaining a constant temperature such as 75 °C, 100 °C and 150 °C. The results were very impressive towards the effectiveness and durability of the coatings.

Keywords: Latex based coatings · Water absorption · Abrasion
Heat resistance · Chloride ion penetration

1 Introduction

The durability of the concrete mainly depends on its quality and a good quality concrete is definitely a durable one. The durability can generally be increased by proper selection of materials, proportioning, proper placing and proper curing of the structure. Typically concrete with higher strength and lower permeability is more durable. Alternatively, concrete can be made durable by the application of coatings. Concrete coatings can provide protection against corrosion of reinforcement due to carbonation, chloride ingress, resistance to chemical attack besides decoration, dust reduction, water

M. Barman et al. (eds.), *Transportation and Geotechniques: Materials, Sustainability and Climate*, Sustainable Civil Infrastructures,
https://doi.org/10.1007/978-3-319-95768-5_12

proofing, improved surface properties, enhanced slip resistance, cleaning ability, etc. Also it protects concrete from damages caused by penetration of moisture, water, heat, abrasion, etc. In addition, coatings provide the first line of defense for concrete from deterioration due to alkali-silica reaction, leaching, thermal and moisture movements, carbonation, chloride ingression, sulphate attack and also against cracking.

There are various materials available for coating on concrete surfaces. Some of them are: cementitious coatings, epoxies, epoxy phenolics, epoxy polyesters, aggregate filled epoxies, silicones, silanes, chlorinated rubbers, thick filled elastomers, etc. These coatings are suggested for certain service conditions such as severe chemical, moderate to severe physical conditions, water repellency, clean ability, aesthetics etc. Also to repair concrete in aggressive environments an elastomeric membrane coated with inorganic cement can be mechanically added to the substrate with studs.

In this paper, Latex was chosen as binder. Fifteen types of coatings were prepared and applied on concrete specimen and evaluated for best performance. The scope of this work is (i) to study the behavior of the concrete coatings and (ii) to evaluate the resistance of these coatings against various conditions so as to identify the best coatings for concrete surfaces.

2 Polymer Coatings for Concrete

Barbucci et al. (1997) [1] compared the water and water vapour permeability coefficients of different concrete coatings such as epoxy, epoxy (water based under coat) and Acrylic-Polyurethane (top coat), epoxy (modified solvented primer) & epoxy (modified solvent free top coat), epoxy (modified, solvent primer) & epoxy polyurethane (top coat) and acrylic (solvented). They concluded that by adopting new methods, the theoretical results were in agreement with that of experimental results. Rodrigues et al. (2000) [2] investigated chloride concentration profiles (Co) and water permeability coefficients by applying acrylic coatings on concrete specimen and the results were compared with control ones. They reported that depending upon formulation, Co was reduced by more than 80%. The ranking of effectiveness to stop chlorides is the same as water imperviousness but the minimum requirement (0.1 Kgm-2 h-0.5) did not prove to be enough for effective protection against chloride penetration. Almusallam et al. (2003) [3] evaluated the durability of concrete coated with five generic type coatings by assessing water absorption, chloride permeability, chloride diffusion and chemical resistance by immersing uncoated and coated samples in 2.5% sulphuric acid. They pointed out that epoxy and polyurethane coatings had a better performance than acrylic, polymer and chlorinated rubber coatings. They noticed that there was a variation in performance of same generic type produced from different manufacturers and the selection warranted trial tests. Yang et al. (2005) [4] estimated the chlorine diffusion into Plain Cement Concrete, fly ash concrete and slag with various water-binder ratios of 0.35, 0.45, 0.55 and 0.65 by conducting ponding tests. Their results proved that for the same w/b ratio, migration of chloride diffusion coefficient of concrete with mineral admixtures was less. Also they suggested that, the diffusion coefficient measured by 180-day ponding test was higher than migration coefficient. Sahmanan et al. (2007) [5] investigated chlorine ion transport properties of Engineered Cementitious Composites

(ECC) by using immersion and salt ponding tests. Their results showed that ECC was effective in slowing down the diffusion of chlorine ion under combined mechanical and environmental loading by its ability of self-controlled tight crack width.

Aguiar et al. (2008) [6] studied the effect of two polymers (acrylic and epoxy) applied over concrete specimen by conducting chloride penetration, sulfates, acids and bases attack tests. They concluded that overall performance of epoxy resin was better than that of other used coatings. Scarfato et al. (2011) [7] tested the effectiveness of nano composite systems at 2, 4, 6% by weight of nanoclay. By applying nano composite systems on concrete substrates they compared the performance under salt attack resistance, porosity, surface water repellency tests and color changing with that of plain resins (Fluoline CP & Anti-pluviol s). The results indicated that nanoclay addition significantly improved protection effectiveness. Shi et al. (2012) [8] explored the feasibility of using modified styrene acrylic emulsion as surface coating on concrete. They inferred that the thicker the coating, the more reduction in mortar shrinkage and enhanced strength at early ages. Also the carbonation resistance, mortar infiltration resistance and chloride diffusion resistance were improved and capillary absorption ratio was reduced by 87% and 78% in dry and standard curing conditions. Zafeiropoulou et al. (2013) [9] experimentally investigated the anticorrosion properties of nine coating systems. They used three different categories of coatings viz: conventional coatings (acrylic paint, elastomeric & acrylic resin dispersion and silicone-acrylic paint), high performance coatings (epoxy, polyurethane and chlorinated rubber) and nanotechnology paint system (silicone coating, pure acrylic paint and an elastomeric-nano acrylic coating). They concluded that the polyurethane coating exhibited good results against chloride induced corrosion and high performance coatings were harmful to environment due to the presence of organic solvents. Also, nano coating I showed improvement regarding chloride-ion corrosion but worst regarding carbonation.

Dang et al. (2014) [10] investigated the effect of three concrete sealers, two crack sealants and two water repellents against salt scaling. Concrete cylinders were subjected to wet/dry cycles in exposure of diluted deicer simulated by 3% by weight of NaCl solution and investigated water absorption rates, gas permeability and water content angle. They confirmed that all the coatings were outstanding among which epoxy based sealer T48CS and water repellent ATS-42 exhibited best performance against salt scaling. Li et al. (2015) [11] determined chloride resistance of concrete specimen applied with typical coatings viz. Epoxy Glass Flake Paint (EP), Poly Urethane paint (PO), Cement Based Permeable Crystallization Waterproof Coating (CE) and Silane Based Water Repellant Coatings (SI) coatings. They conducted studies under outdoor natural climate conditions and indoor artificial accelerated experiments using ultra violet light radiation and wetting/drying cycles. The results showed that chloride resistance was remarkably improved, organic film coatings deteriorate faster than infiltrating coatings and service lives of concrete coatings were closely related with solar irradiance. It is clear from the above studies that most of the coatings were solvent based in protecting concrete in order to improve durability and very little studies were carried out on the durability of concrete using latex based coatings. Therefore, an attempt is made to study the performance of concrete provided with latex based coatings.

3 Materials and Methods

Materials used

For this study, water proofing latex purchased from Madurai (India), was used as binder. The functional filler materials such as fly ash, Ordinary Portland Cement (OPC), Feldspar, High Alumina Clay, Kaolin, Micro silica, Black Clay, Zinc phosphate were purchased locally. The chemical components like Aluminium Oxide ($Al_2O_{3)}$, Titanium Di Oxide ($TiO_{2)}$, Silica Fume were procured from Himeda (India). The concrete specimen used for various tests were: Big Tiles of size 230 × 230 × 20 mm, Small Tiles of size 100 mm × 100 mm × 10 mm, disc of size 80 mm diameter and 40 mm length and small cylinder of size 100 mm length and 40 mm diameter, which are shown in Fig. 1. These specimen were prepared by mixing cement with sand in the ratio of 1:3 by maintaining a water cement ratio of 0.45.

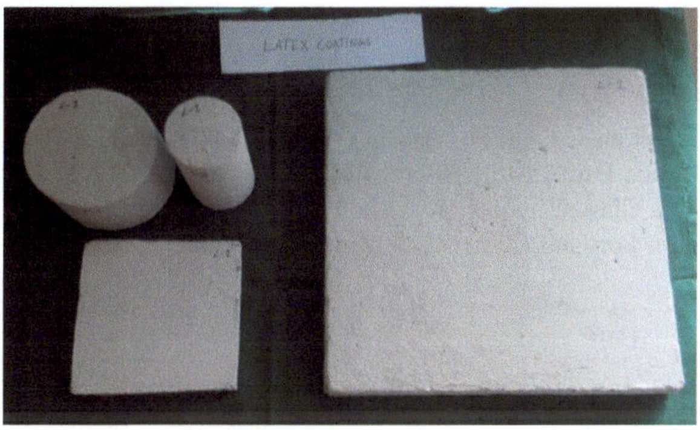

Fig. 1. Test specimen with latex coating

Test specimen

The cylindrical specimen of size 100 mm length and 40 mm diameter were utilized to evaluate the water absorption capacity of the concrete coating. Small Tiles of size 100 × 100 × 10 mm were used to determine abrasion resisting capacity of the coating. The circular disc of 80 mm diameter and 40 mm length were used to determine the chloride ion penetration into the coated specimen. Big Tiles of size 230 × 230 × 20 mm were used to determine the heat resisting capacity of the coating.

Coatings used

The different combinations of coating systems investigated in this study are presented in Table 1. After the concrete specimen have been cured for 28 days and dried for several days in the open space (until it is completely dry), the surface was applied with the coatings twice (the second coating was applied after the first coating was completely dry).

Table 1. Compositions of different coating systems

Coating designation	Materials used
LS1	L + FA + A +T + SF
LS2	L + MS + A +T + SF
LS3	L + F + A +T + SF
LS4	L + HAC + A +T + SF
LS5	L + FA + MS + A +T + SF
LS6	L + FA + Z + A +T + SF
LS7	L + FA + C + A +T + SF
LS8	L + FA + MS + C + Z + A +T + SF
LS9	L + FA + MS + C + A +T + SF
LS10	L + FA + F + A +T + SF
LS11	L + FA + K + A +T + SF
LS12	L + BC + A +T + SF
LS13	L + K + A +T + SF
LS14	L + OPC + FA + F + HAC + A +T + SF
LS15	L + OPC + MS + F + HAC + A +T + SF

Note L—Latex, FA—Fly Ash, A—Alumina, T—Titanium dioxide, SF—Silica Fume, MS—Micro Silica, F—Feldspar, HAC—High Alumina Clay, C—Clay, K—Kaolin, BC—Black Clay, OPC—Ordinary Portland Cement, Z—Zinc phosphate

4 Test Analysis

1. **Water Absorption Test**

Cylindrical specimen of size 40 mm diameter and 100 mm length were used for the determination of water absorption. The weight of uncoated and coated specimen were measured before and after immersion in water for a period of 48 h and the amount of water absorbed was noted using the expression given in Eq. (1)

$$\text{Percentage of Water Absorption} = (\text{B} - \text{A}/\text{A}) \times 100 \tag{1}$$

where,

A = Initial weight of the Coated sample
B = Weight of the saturated sample

Water absorption for all the coatings were less than that of the control one. Further, water absorption for LS4, LS6 & LS15 was comparatively much less i.e. less than 1.2%. Out of which the combination of LS15 showed much better impermeability and the water absorption of LS15 was 0.65% only. Lower the water absorption, higher the durability of concrete. So, concrete with coating LS15 exhibited good impermeability (Fig. 2, Table 2).

Fig. 2. Water absorption test

Table 2. Test for determining the water absorption capacity of the specimen

Sl. No	Designation	Dry weight of the specimen (g)	Weight of the coated specimen (g)	Wet weight of the specimen (g)	Water absorption (%)
1	**Control**	283.92	–	294.65	3.78
2	**LS1**	288.01	290.14	299.22	3.13
3	**LS2**	286.12	294.02	298.92	1.67
4	**LS3**	270.18	274.12	284.72	3.87
5	**LS4**	292.36	299.82	303.40	1.19
6	**LS5**	291.03	296.23	302.13	1.99
7	**LS6**	290.23	300.79	303.96	1.05
8	**LS7**	284.17	286.20	289.91	1.30
9	**LS8**	293.63	297.13	301.50	1.47
10	**LS9**	286.27	295.77	300.78	1.69
11	**LS10**	290.55	299.21	303.02	1.27
12	**LS11**	285.23	289.03	295.03	2.07
13	**LS12**	289.37	294.38	304.06	3.29
14	**LS13**	292.57	300.04	304.25	1.40
15	**LS14**	294.72	305.23	309.41	1.37
16	**LS15**	**283.92**	**291.19**	**293.09**	**0.65**

2. **Abrasion Test**

The specimen of size 100 mm × 100 mm × 10 mm was used for abrasion test. Initially, the weight of coated specimen were measured and later they were scribed ten times by using E4 320 emery sheet with uniform thrust. After scribing, the specimen were weighed to determine the loss of weight of the coating.

If the loss of weight is less, it would mean that the capacity of the coating against abrasion is better. Here, the coatings LS1, LS6, LS7, LS12 and LS15 proved that the loss of weight was only 10 mg. This showed that these coatings exhibited good resistance against abrasion (Table 3).

Table 3. Testing for the abrasion resistance of the specimen

Sl. No	Designation	Initial weight (g)	Final weight (g)	Loss of weight (g)	Loss of weight (mg)
1	**LS1**	274.10	274.00	0.1	10
2	**LS2**	298.88	298.82	0.06	60
3	**LS3**	292.20	292.18	0.20	20
4	**LS4**	265.80	265.76	0.04	40
5	**LS5**	251.65	251.62	0.03	30
6	**LS6**	262.01	262.00	0.01	10
7	**LS7**	234.50	234.49	0.01	10
8	**LS8**	285.70	285.67	0.03	30
9	**LS9**	304.02	303.98	0.04	40
10	**LS10**	247.93	247.91	0.02	20
11	**LS11**	258.21	258.18	0.03	30
12	**LS12**	282.33	282.32	0.01	10
13	**LS13**	249.82	249.80	0.02	20
14	**LS14**	244.02	243.95	0.07	70
15	**LS15**	311.82	311.81	0.01	10

3. **Chloride Ion Penetration Test (ASTM C 1202)**

The chloride ion penetration test covers the determination of the electrical conductance of concrete to provide a rapid indication of its resistance to the penetration of chloride ions. The test method consists of monitoring the amount of electrical current passed through 40 mm length and 80 mm diameter disc specimen during 6 h period. A potential difference of 60 V direct current was maintained across the ends of the specimen, one of which is immersed in a 3% Sodium Chloride (NaC1) solution and the other in 0.1 M Sodium Hydroxide (NaOH) solution. Figure 3 shows the test setup for Chloride ion penetration test.

The total charge passed in coulombs, had been found to be related to the resistance of the specimen to chloride ion penetration. The cell containing 3% NaC1 solution was connected to the negative terminal of the power supply, the other cell containing 0.1 Molarity NaOH was connected to the positive terminal of the power supply. After

Fig. 3. Chloride ion penetration test setup

fixing the specimen within the cell, solutions are filled in the cell and 60 V direct current was applied. Once the power supply is turned on, the initial current is measured with the ammeter. After every 30 min till the completion of 6 h, the current was monitored and recorded in each cell. The charge passed was computed by using trapezoidal rule. The charge passed is expressed in coulombs (1 C is 1 A second).

$$\mathbf{Q = 900[I_0 + 2(I_{30} + I_{60} + \ldots\ldots\ldots\ldots I_{300}) + I_{360}]} \tag{2}$$

where
Q Charge Passed (coulombs)
I_0 Current (amperes) immediately after voltage was applied
I_T Current (amperes) at t minute after voltage was applied.

The measured charge was corrected by multiplying the value obtained from the above ratio of the cross sectional area of the standard and actual specimen i.e.

$$\mathbf{Q_S = Q\,x\,(3.75/X)^2} \tag{3}$$

where,
Q_S Charge passed (coulombs) through 3.75 in. (95 mm) diameter
Q Charge passed (coulombs) through y inch diameter specimen
X Diameter of the standard specimen.

Depending upon the charge passed through the specimen, the permeability characteristics could be stated. Table 4 gives the relationship between charge passed and chloride ion permeability. The results of chloride ion permeability tests are given in Table 5.

Table 4. Chloride ion penetration standards as per ASTM C 1202

Sl. No	Charge passed	Chloride ion permeability
1	>4000 C	High
2	2000–4000 C	Moderate
3	1000–2000 C	Low
4	100–1000 C	Very Low
5	<100 C	Negligible

Higher resistance to Chloride ion penetration represents a good coating. The chloride ion diffusivity was much less for almost all of the combinations except LS2 when compared to the control specimen. For coatings LS8, LS9, LS13, diffusion of chloride ions was low and for LS14 and LS15 was comparatively much less having

Table 5. Chloride ion penetration test results (ASTM C 1202)

Designation	Charge passed in coulombs	Chloride ion permeability	Designation	Charge passed in coulombs	Chloride ion permeability
Control	3602.47	Low	**LS8**	1326.52	Low
LS1	369.42	Very low	**LS9**	1962.96	Low
LS2	6033.29	High	**LS10**	333.27	Very low
LS3	823.63	Very low	**LS11**	397.31	Very low
LS4	801.46	Very low	**LS12**	296.57	Very low
LS5	778.65	Very low	**LS13**	1755.22	Low
LS6	514.21	Very low	**LS14**	224.02	Very low
LS7	454.83	Very low	**LS15**	215.92	Very low

charge passed as 224.02 and 215.92 C respectively. Concrete with Coating LS15 proved to be an excellent coating.

4. **Heat Resistance Test**

This test was carried out in a muffle Furnace with top loading attached with temperature Indicator and thermocouple as shown in Fig. 4. This test was carried out for 230 × 230 × 20 mm thick tiles for all coated mixes. Constant temperature ranges such as 75 °C, 100 °C and 150 °C were arbitrarily chosen for this study. The tile to be tested was placed horizontally over the top of the furnace and the temperature was set by adding the knob, located in digital temperature indicator. Initially the set temperature of 75 °C was maintained for a period of 30 min. The bottom phase of the tile was exposed to the set temperature for 30 min. Using laser Torch thermometer, the temperature of the bottom surface and top surface were measured and recorded. The experiment was continued for 100 °C and 150 °C set temperature with a soaking period of 15 min. The corresponding temperatures and temperature drops were

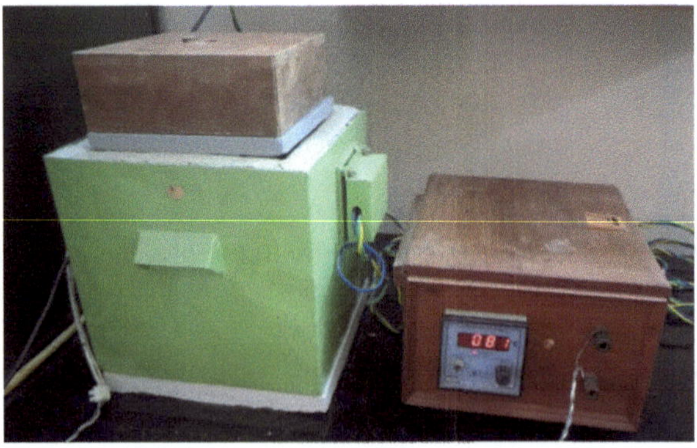

Fig. 4. Tiles placed in muffle furnace

Table 6. Testing for the heat resistance of the coated tiles

Sl. No	Designation	30 min (75 °C)		15 min (100 °C)		15 min (150 °C)		Swing for 75 °C	Swing for 100 °C	Swing for 150 °C
		Top	Bottom	Top	Bottom	Top	Bottom			
1	**Control**	57.5	58.4	68.4	71.5	99.1	105.3	0.9	3.1	6.2
2	**LS1**	55.8	60.2	64.1	69.3	88.8	96.4	4.4	5.2	7.6
3	**LS2**	51.9	57.2	69.2	75.0	92.3	99.0	5.3	5.8	6.7
4	**LS3**	52.8	57.5	59.8	65.1	73.9	87.1	4.7	5.3	13.2
5	**LS4**	58.1	62.7	71.8	79.6	93.2	100.2	4.6	7.8	7.0
6	**LS5**	52.5	55.3	56.6	59.1	75.3	85.5	2.8	2.5	10.2
7	**LS6**	69.8	74.2	75.3	79.4	84.0	89.1	4.4	4.1	5.1
8	**LS7**	63.0	67.9	69.9	76.3	80.2	91.9	4.9	6.4	11.7
9	**LS8**	52.6	55.1	64.7	73.2	81.3	92.6	2.5	8.5	11.3
10	**LS9**	53.2	56.6	64.8	70.3	75.9	89.7	3.4	5.5	13.8
11	**LS10**	58.4	61.2	59.9	65.2	74.8	84.0	2.8	5.3	9.2
12	**LS11**	53.5	56.8	61.8	67.8	79.5	90.0	3.3	6.0	10.5
13	**LS12**	60	62.9	70.6	75.2	86.2	102.2	2.8	4.4	16.0
14	**LS13**	63.8	66.4	74.3	77.3	88.6	95.1	2.6	3.0	6.5
15	**LS14**	53.7	55.6	64.5	68.0	83.8	86.4	1.9	3.5	2.6
16	**LS15**	56.4	61.5	65.5	74.3	91.9	110.0	5.1	8.8	18.1

recorded separately for all categories of coated tiles and the results were presented in Table 6.

It was observed from Table 6 that, combinations LS3, LS9, LS12 and LS15 showed higher temperature difference between the top and bottom surfaces of the tiles which implied that the coatings much resisted the passage of heat inside the tiles. Among the above, coating LS15 showed a much better resistance to heat for all the 3 swings. Moreover there was a vast variation between the top and bottom temperatures of swing for 150°.

5 Conclusions

On the basis of the research work carried out, the following conclusions are drawn.

1. Water absorption test was carried out for all the specimen including control specimen. Control specimen exhibited water absorption value of 3.78%. The water absorption value of the coating LS15 was 0.65% which is the least and it is 5.82 times better than that of the control specimen. All the other coatings showed higher water absorption rate than LS15.
2. In abrasion resistance test, all the coated specimen were tested and loss of weight of coatings was observed. A least value of 10 mg was observed for LS1, LS6, LS7, LS12 and LS15 and the poor value of 70 mg was observed for LS14. This is due to

the presence of Fly Ash with Ordinary Portland Cement, which has not undergone pozzolanic reaction.

3. Chloride ion penetration test as per ASTM C 1202 had been carried out on all the specimen and LS15 showed 215.92 C of charge passed and this is due to the presence of Ordinary Portland Cement in LS15. The enhanced film tightness had been obtained by the pozzolanic activity of Micro Silica with Ordinary Portland Cement and therefore LS15 was found to be the best than all other coatings. Highest value of 6033.29 C was obtained for LS2 and this was due to the absence of Ordinary Portland Cement in LS2.
4. In Heat Resistance Test, LS15 was found to be the best among all the coatings with temperature swings of 5.1, 8.8 and 18.1 for the temperatures of 75 °C, 100 °C and 150 °C with a soaking period of 30 min, 15 min and 15 min respectively.

References

1. Barbucci, A., Delucchi, M., Cerisola, G.: Organic coatings for concrete protection: liquid water and water vapour permeability. In: Proceedings, ELSEVIER Science Progress in Organic Coatings. pp. 293–297 (2000)
2. Rodrigues, M.P.M.C., Costa, M.R.N., Mendes, A.M., Eusebio M.: Effectiveness of surface coatings to protect reinforced concrete in marine environment. Sci. Rep. Mater. Struct. **33**, 618–626 (2000)
3. Almusallam, A.A., Khan, F.M., Dulaijan, S.U., Al-Amoudi, O.S.B.: Effectiveness of surface coatings in improving concrete durability. Cement Concr Comp **25**, 473–481 (2003). (Elsevier)
4. Yang, C.C., Lin, S.S. Cho., S.W.: Relation between migration coefficient from accelerated chloride migration test and diffusion coefficient from ponding test. Kuwit J. Sci. Eng. **32**(2), 165–186 (2005)
5. Sahmanan, M., Li, M., Li, C.: Transport properties of engineered cementitious composition under chloride exposure. ACI Mater. J. 303–310 (2007)
6. Aguiar, J.B., Cmoes, A., Moreira, P.M.: Coatings for concrete protection against aggressive environment. J. Advan. Concr. Technol. **6**(1), 243–250 (2008)
7. Scarfato, P., Di Maio, L., Fairello, M.L., Russo, P., Incarnato, L.: Preparation and evaluation of polymer/clay naonocomposite surface treatments for concrete durability enhancement. Cement Concr. Comp. **34**, 297–305 (2011). (Elsevier)
8. Shi, L., Liu, J., Liu, J.: Effect of polymer coating on the properties of surface layer concrete. Proc. Eng. **27**, 291–300 (2012). (Elsevier)
9. Zafeiropoulou, Theodosia, Rakanta, Eleni, Batis, George: Carbonation resistance and anticorrosive properties of organic coatings for concrete structures. J. Surf. Eng. Mater. Advan. Technol. **2013**(3), 67–74 (2013). https://doi.org/10.4236/jsemat.2013.31A010
10. Dang, Y., Shi, X., Mery, S., Xie, N., Benson, A., Yang, Z.: Influence of surface sealers on the properties of internally cured cement mortars containing saturated fine lightweight aggregate. J. Mater. Civil Eng. **12** (2015)
11. Li, G., Yang, B., Guo, C., Du, J., Wu, X.: Time dependence and service life prediction of chloride resistance of concrete coatings. Constr. Build. Mater. **83**, 19–25 (2015). https://doi.org/10.1016/j.conbuildmat.2015.03.003

One Dimensional Ground Response Analysis and Identification of Liquefiable Strata of Guwahati City

Binu Sharma(✉), Amar F. Siddique, and Bhaskar J. Medhi

Civil Engineering Department, Assam Engineering College, Guwahati, India
{binusharma78, amaribnesiddiq98, bhaskar08jec}
@gmail.com

Abstract. According to seismologist the Himalayan region between the epicenter of the 1897 great Assam earthquake and the 1950 Assam earthquake is due for a large earthquake of magnitude greater than 8. Guwahati city lies at about 100–200 km from this zone. This has highlighted the need to better understand the vulnerability of land to seismic hazards caused by future earthquakes in Guwahati city. The effect of large modifications in seismic waves that occur due to variation in soil properties near the surface of the earth is of great importance in geotechnical earthquake engineering. The response of a structure depends on its regional seismicity, source mechanism, geology and local soil conditions. This work is based on the one dimensional ground response analysis (GRA) using the equivalent linear technique conducted for one hundred twenty (120) locations in Guwahati city. The input motion of 2011 Sikkim earthquake (M_w = 6.9) having bedrock PGA of 0.152 g at 30 m depth has been used for the purpose. This motion is selected because it belongs to the same tectonic regime. The results of variation of PGA with depth, strain and maximum stress ratio distribution over depth are obtained for all boreholes. It has been observed that the peak ground acceleration (PGA) of the input motion gets amplified at ground surface. The surface PGA was found to vary from a maximum value of 0.62 g at Kalapahar (in Paltanbazar area) to a minimum value of 0.15 g near Bar Sajai Ghoramara (in Beltola area) in Guwahati city. The spectral accelerations at various locations have been compared with the spectral acceleration of rocky or hard soil sites (IS: 1893–2002) for the bedrock 0.152 g PGA input motion. Deterministic assessment of SPT based liquefaction potential was next carried out. The surface PGA was used to determine the cyclic stress ratio in each location. Liquefaction susceptibility as determined is presented as a GIS based map showing zones of levels of risk of liquefaction. This work has been mainly carried out to initiate such studies in this seismic prone region and to address the problems caused due to seismic hazards.

M. Barman et al. (eds.), *Transportation and Geotechniques: Materials, Sustainability and Climate*, Sustainable Civil Infrastructures,
https://doi.org/10.1007/978-3-319-95768-5_13

1 Introduction

North Eastern region of India, which is one of the six most seismically active regions of the world, lies at the junction of the Himalayan arc to the north and the Burmese arc to the east. Guwahati which lies in the North Eastern region between coordinates 26.1833N and 91.733E is situated on the bank of the mighty Brahmaputra, surrounded by hills from the southern and eastern sides of the city. The city falls in the highest seismic risk zonal level i.e. zone V as per IS 1893 (2002). According to Nath et al. (2008), the city had been hit by earthquakes of different magnitudes ranging from 5 to 8.7 in the past. This has highlighted the need to better understand the vulnerability of land to seismic hazards caused by future earthquakes in Guwahati city. Guwahati soil consists of alluvial deposits with layers of both coarse and fine grained soils. Hence, large modifications in earthquake waves can occur due to change in variation in soil properties near to the surface of earth. Clayey sites amplify about 1.5 times higher than the rocky and sandy sites (Boominathan et al. 2008). Ground response analyses are used to predict ground surface motions for development of design response spectra, to evaluate dynamic stresses and strains, to identify liquefaction hazards and to determine the earthquake-induced forces that can lead to instability of earth and earth-retaining structures. Due to the high seismicity level together with presence of soft alluvial deposits in Guwahati city, one dimensional ground response analysis (GRA) using the equivalent linear technique was conducted for one hundred twenty (120) locations in Guwahati city. The input motion of 2011 Sikkim earthquake ($M_w = 6.9$) having bedrock PGA of 0.152 g at 30 m depth has been used for the purpose. This motion is selected because it belongs to the same tectonic regime of North East India. It is also important to consider the effect of local geology on liquefaction potential analysis. There are very few works on liquefaction potential of Guwahati soil considering the local site effects. Raghukanth and Dash (2009) estimated the liquefaction potential of Guwahati city taking the 1869 Cachar earthquake (Mw = 7.6), 1897 great Assam earthquake (Mw = 8.7) and a probable future earthquake in Assam gap (Mw = 8.5). The authors determined the rock level peak ground acceleration (PGA) using finite fault seismological model of Motazedian and Atinkinson (2005) and the model was implemented in the time domain through attenuation relations suggested by Boore (1983). This work reports the 1D ground response analysis (GRA) for one hundred twenty (120) boreholes in Guwahati city using the 2011 Sikkim motion earthquake. It is also intended to evaluate the liquefaction potential of Guwahati city considering the response of the soil deposit pertaining to bedrock PGA of 0.152 g of the Sikkim motion earthquake.

2 Methodology

For one dimensional ground response analysis three methods are used i.e. the Linear, Equivalent linear and Non-linear method. The equivalent linear method of ground response analysis was developed to analyze the nonlinear response of soil using frequency domain analysis with the aid of linear transfer functions. The non-linear behaviour of the soil is modeled in terms of equivalent linear properties corresponding to effective shear strain using iterative procedure (Kramer 1996).

In this work the one dimensional equivalent linear method is used to perform ground response analysis using the Deepsoil software. The Deepsoil software has been developed to perform one dimensional ground response analyses (Hashash et al. 2011). For this purpose, an input motion data is required which can be obtained from the seismograph stations. The input motion of 2011 Sikkim earthquake which is recorded in the Gangtok station ($M_w = 6.9$) having bedrock PGA of 0.152 g at 30 m depth has been used for the purpose. The details of the characteristics of the motion are given in Table 1. The recorded data is in a rocky site falling in site class A and having a density of 25kN/m^3 and shear wave velocity of 1500 m/s. Subsurface data necessary for the work are taken from a data base of one hundred twenty (120) boreholes upto 30 m depth in Guwahati city. The subsurface data consists of bore hole stratifications, standard penetration test (SPT) N values and engineering properties of the soils and depth of water table. Shear wave velocity is one of the most important input parameters to represent the stiffness of the soil layers. Deepsoil allows the alternative input of shear wave velocity in place of shear stiffness. The shear wave velocity of the boreholes at each depth which is needed for the analysis is calculated from Sharma and Rahman (2016) using the Eq. (1).

Table 1. Strong motion parameters for 2011 Sikkim earthquake

Strong motion parameters	2011 Sikkim earthquake
Magnitude	6.9
Recorded station	Gangtok
Site class	A
Epicentral distance	68 km
Bedrock PGA (g)	0.152
Predominant period (sec)	0.14
Bracketed duration (sec)	73.90
Significant duration (sec)	31.60
Arias intensity (m/sec)	0.665
Specific energy density (cm^2/sec)	213.83

$$Vs = 74.639 * N^{0.3876} \quad (1)$$

where, Vs = shear wave velocity in m/sec and N = uncorrected SPT-N value.

SPT based deterministic approach of Idriss and Boulanger (2004) is used for analyzing liquefaction potential.

Material Curves

For defining the strain dependent shear modulus and damping ratio for soil layers a number of curves have been defined in Deepsoil software. When the modulus reduction (for shear modulus) and damping curves for a site is absent, the curves proposed by Vucetic and Dobry (1991) for clay and Seed and Idriss (1970) for sandy soil are considered to be better alternatives (Hashash et al. 2011). Hence, for this study these two curves have been used to define the dynamic properties of soil layers. For sands, effective vertical stress is required for defining the modulus reduction curve. In case of clay, an addition of plasticity index is needed with vertical effective stress for defining modulus reduction and damping curves.

3 Soil Study of Guwahati City

The study area in Guwahati city, measuring approximately about 150 km^2, covers the southern and eastern part of the city. The city has a population of over one million (Census 2011), occupying mostly the narrow tracts of alluvium and sediment filled low lands interspersed with Precambrian residual hills. Alluvial soil is found in the valleys and low lying areas of the city. They are typically brown and grey coloured silty clays or clayey silts. The hills surrounding Guwahati city are primarily composed of Porphyritic Granites and Quartzo Feldspathic Gneiss which are cross-cut by amphibolite intrusives and quartz veins. Sandy soils which are produced by weathering of porphyritic granites are found in many areas of the city. These soils are pale brown in colour and have a low degree of cohesion.

Data of one hundred twenty (120) bore holes upto 30 m depth were taken from a project funded by the Department of Science and Technology, India. From engineering point of view, soils in Guwahati city mostly consists of alternating layers of both fined grained and coarse grained soils of varying thickness. The fine grained fraction mostly consists of soils of classification CL, CI and CH according to the Indian standard soil classification. In a few locations inorganic silt of classification ML and CL-ML and non plastic inorganic silts were also encountered. The coarse grained fraction is mostly of classification SP, SW, SC, SM, SP-SC. Gravel deposits were also encountered in certain boreholes.

4 Results and Discussion

The ground motion of the Sikkim earthquake is used for carrying out the one dimensional equivalent linear ground response analysis (EQL GRA) for the one hundred twenty (120) boreholes using Deepsoil software. Rock strata are not encountered upto 30 m in the one hundred twenty (120) bore holes. It has been found that the top 30 m of soil stratum greatly influence the ground motion characteristics and site amplifications. Hence using the bedrock PGA of 0.152 g at 30 m depth of the Sikkim earthquake, the PGA at different depths of the one hundred twenty (120) boreholes is determined. The strain and maximum stress ratio distribution over depth are also obtained for all boreholes. It has been observed that the bedrock PGA of the input motion gets amplified at ground surface. The range of surface PGA and the range of amplification factors of sixteen sites comprising the one hundred twenty (120) bore holes are shown in Table 2. The soft clayey sites have been found to highly amplify the bedrock PGA. The surface PGA was found to vary from a maximum value of 0.62 g at Kalapahar (in Paltanbazar area) to a minimum value of 0.15 g near Bar Sajai Ghoramara (in Beltola area) in the eastern and the southern part of Guwahati city. Strong amplification of seismic motions are found to occur in Fatashil Ambari, Jonali, Kachari, Lal Ganesh, Paltanbazar and Ulubari locations whereas Hatigaon and Guwahati club areas showed less amplification due to the presence of deeper sandy deposits. The Fourier amplitude ratio (FAR) at surface level for the southern and eastern part of Guwahati city has been obtained in the range of 3.75-9.49 with fundamental frequency range from 1.02 Hz to 3.75 Hz. This is in agreement with the FAR at surface level for Guwahati city from Kumar and Krishna (2012) who have obtained FAR in the range of 2.46–10.0 with fundamental frequency range from 2.9 Hz to 4.14 Hz. The range of FAR values and the fundamental frequency in Hz for the sixteen sites are shown in Table 2. Also the variation of maximum shear strain values for the above mentioned sixteen sites are shown in the same table.

Table 2. Range of responses for all sites having 120 Boreholes (S PGA = Surface PGA, A.F. = Amplification Factor, FAR = Fourier Amplitude Ratio, F Freq = Fundamental Frequency, PSA = Peak Spectral Acceleration)

Site location	S PGA (g)	A. F.	FAR	F Freq (Hz)	PSA (g)	Max. Strain (%)
Beltola	0.15–0.35	1.00–2.30	4.13–5.93	1.02–2.38	0.52–1.69	0.00–1.10
Bharalumukh	0.32–0.60	2.11–3.95	5.26–6.44	1.38–2.55	1.11–1.99	0.02–0.20
Panbazar	0.30–0.53	1.97–3.49	5.57–6.65	1.67–2.55	1.03–2.06	0.02–0.18
Lal Ganesh	0.29–0.49	1.91–3.22	5.27–7.14	1.67–3.75	1.23–2.22	0.01–0.13
Machkhowa	0.24–0.34	1.58–2.24	4.76–6.32	1.32–2.42	0.88–1.80	0.01–0.25
Chandmari	0.18–0.41	1.18–2.70	3.78–6.00	1.07–2.37	0.63–1.80	0.00–1.58
Bamunimaidan	0.23–0.49	1.51–3.22	4.67–7.57	1.35–2.39	0.91–2.23	0.00–0.71
RGB	0.29–0.37	1.91–2.43	4.97–7.13	1.43–2.05	1.10–1.42	0.01–0.37
Lachit Nagar	0.21–0.56	1.38–3.68	4.45–7.00	1.02–2.79	0.59–3.20	0.01–0.47
Bhangagarh	0.22–0.41	1.45–2.70	4.95–5.48	1.29–2.34	0.70–1.46	0.01–0.15
Six mile	0.36–0.44	2.37–2.89	4.51–5.59	1.50–2.48	1.07–2.51	0.01–0.19
Panjabari	0.22–0.29	1.45–1.91	4.55–5.44	1.05–1.89	0.69–1.49	0.00–0.43
Hatigaon	0.19–0.38	1.25–2.5	3.75–6.04	1.18–2.13	0.65–1.58	0.01–0.51
Paltanbazar	0.31–0.62	2.04–4.08	5.32–9.49	1.94–3.65	1.15–2.76	0.01–0.41
Khanapara	0.23–0.41	1.51–2.70	4.59–5.98	1.16–2.60	0.88–1.93	0.02–0.26
Kahilipara	0.31–0.42	2.04–2.76	4.84–5.66	1.48–2.19	1.00–2.11	0.01–0.19

Using the surface PGA values that have been obtained from ground response analysis, liquefaction potential of southern and eastern part of Guwahati city has been evaluated. Deterministic method of Idriss and Boulanger (2004) has been used to determine the factor of safety against liquefaction. The soil layers that were identified for liquefaction analysis are fine to medium sands, silty sands and clayey sand that have classification of SP, SW, SC, SM, SP-SC. Inorganic silt of classification ML, ML-CL and non plastic inorganic silts were also analyzed for liquefaction susceptibility. Out of the one hundred twenty (120) borehole locations, fifty four (54) boreholes were identified for liquefaction susceptibility and are considered for analysis.

The responses from ground response analysis are discussed further in terms of PGA profile, amplification factor, Fourier amplitude ratio (FAR), maximum stress ratio and maximum strain distribution over depth, spectral acceleration (SA) and liquefaction susceptibility of four arbitrarily selected sites located at Beltola, Panbazar, Lal Ganesh and Bharalumukh area of the sixteen sites using the 2011 Sikkim motion earthquake.

The number of bore holes analysed in the Beltola, Panbazar, Lal Ganesh and Bharalumukh area are 9, 8, 5, and 6 respectively. The surface PGA amplifies from 0.15 g to 0.35 g in the Beltola site, from 0.3 g to 0.53 g in Panbazar site, from 0.29 g to 0.49 g in Lalganesh site and from 0.32 g to 0.60 g in Bharalumukh site. The amplification factor, i.e. the ratio of surface PGA to bedrock acceleration, has been found to be in the range of 1.0–2.30, 1.97–3.49, 1.91–3.22 and 2.11–3.95 in the four sites respectively. The variation of PGA with depth are shown from Figs. 1, 2, 3 and 4. for the four sites.

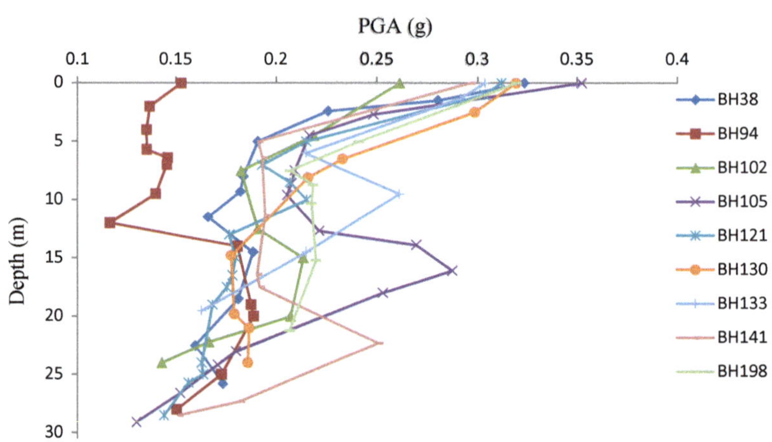

Fig. 1. PGA profile with depth for Beltola area

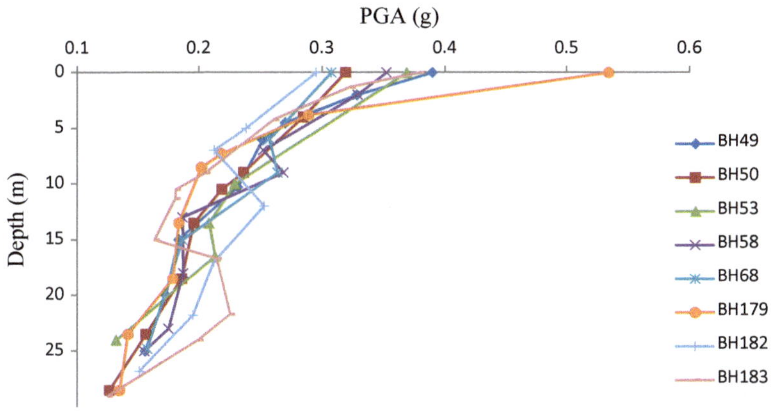

Fig. 2. PGA profile with depth for Panbazar area

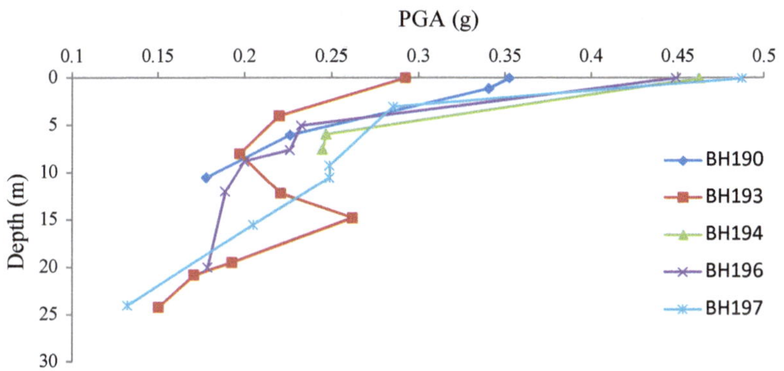

Fig. 3. PGA profile with depth for Lal Ganesh area

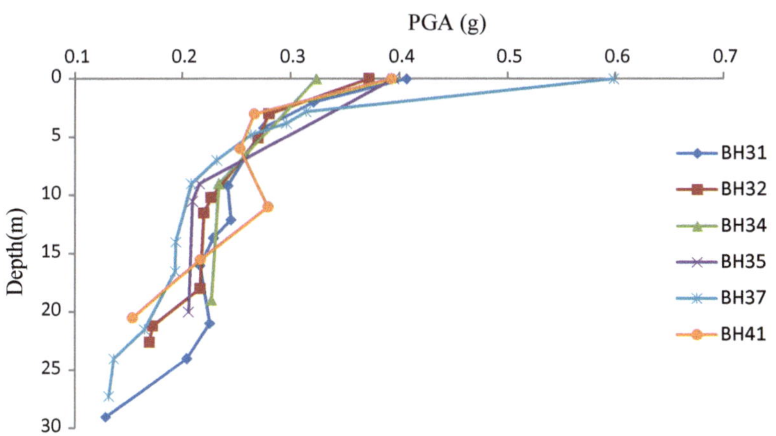

Fig. 4. PGA profile with depth for Bharalumukh area

The Fourier amplitude ratio (FAR) at the surface level in the Beltola area varies from 4.13 to 5.93 with the fundamental frequency range from 1.02 to 2.38 Hz. The FAR in the Panbazar area varies from 5.57 to 6.65 with the fundamental frequency range from 1.67 to 2.55 Hz. Similarly for Lal Ganesh and the Bharalumukh area, the Fourier amplitude ratio at the surface layer varies from 5.27 to 7.14 and from 5.26 to 6.44 with the fundamental frequency range from 1.67 to 3.75 Hz and from 1.38 to 5 Hz respectively. Figures 5, 6, 7 and 8 show the maximum stress ratio distribution with depth, which highlights the effect of local site on different borehole locations corresponding to a single ground motion. The maximum shear strain distribution with depth is presented in Figs. 9, 10, 11 and 12. For less stiffer soils having low SPT-N value or shear wave velocity, higher maximum strain is released. More dissipation of energy is associated with higher strain levels which lead to the softening of the soil with number of loading cycles. Again less dissipation of energy leads to higher amplification of the PGA at various levels of the soil strata. In Beltola area eight borehole sites show a variation of maximum shear strain in the range of 0–0.35%, but the site at B.H.94 shows the highest value of maximum strain of 1.1% at depth 6–7 m due to low value of shear wave velocity. A wide range of maximum strain variation of 0.02–0.18% has been obtained for the Panbazar area. There is a variation of maximum strain of 0.01–0.13% for the Lalganesh area and from 0.02–0.2% in the Bharalumukh area.

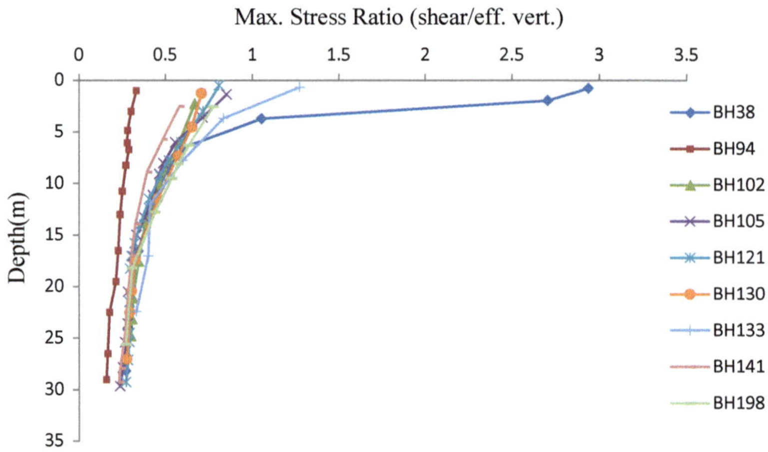

Fig. 5. Maximum shear stress ratio profile at Beltola

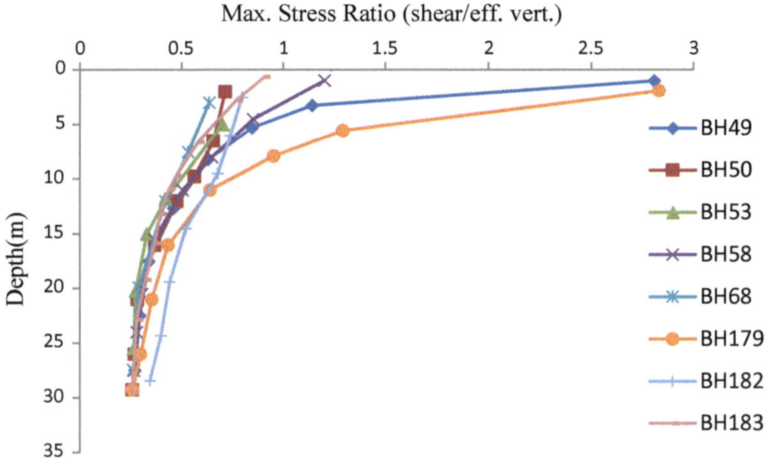

Fig. 6. Maximum shear stress ratio profile at Panbazar

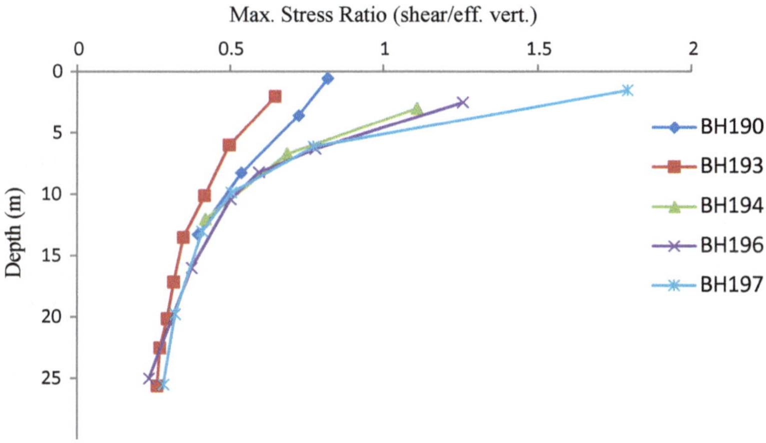

Fig. 7. Maximum shear stress ratio profile at Lal Ganesh

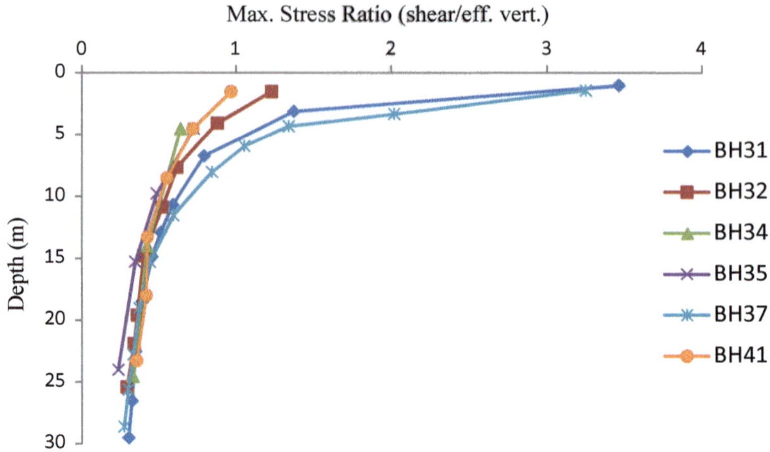

Fig. 8. Maximum shear stress ratio profile at Bharalumukh

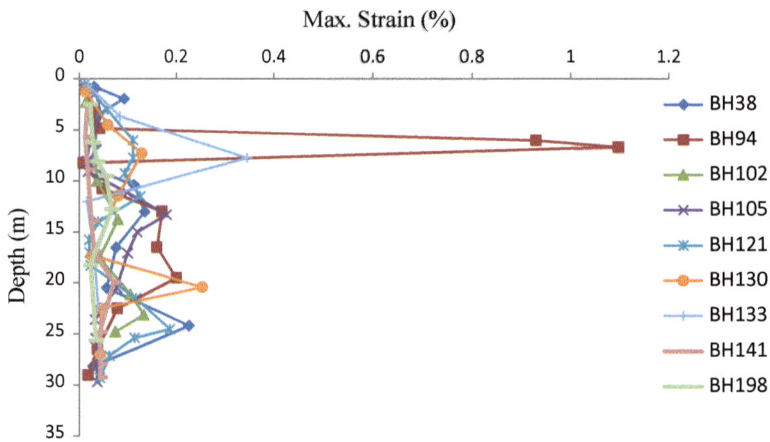

Fig. 9. Maximum strain distribution at Beltola

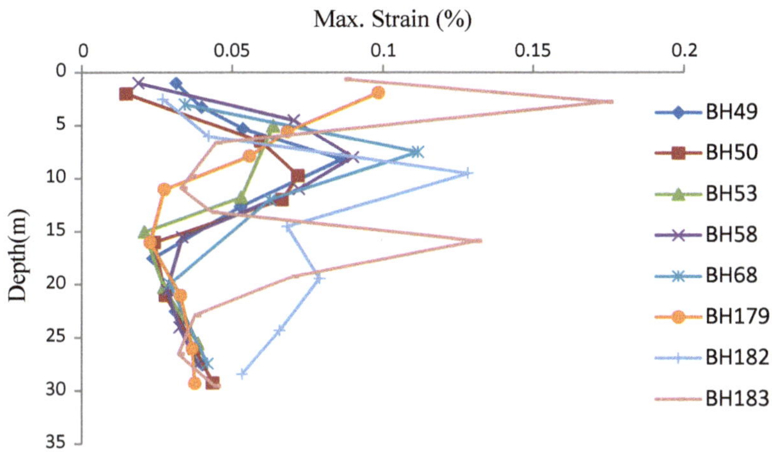

Fig. 10. Maximum strain distribution at Panbazar

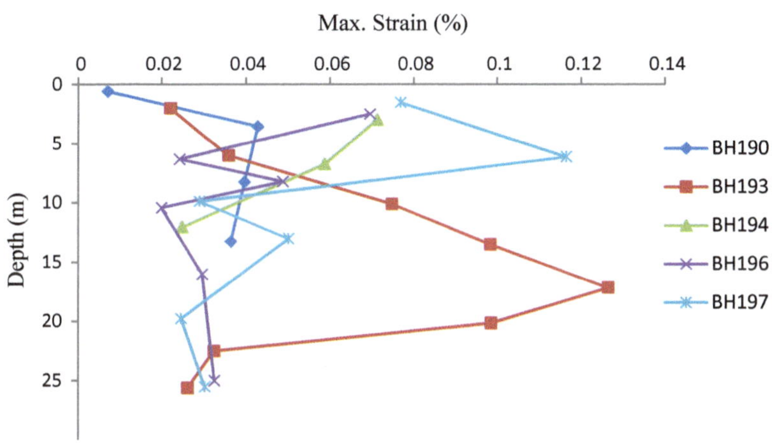

Fig. 11. Maximum strain distribution at Lal Ganesh

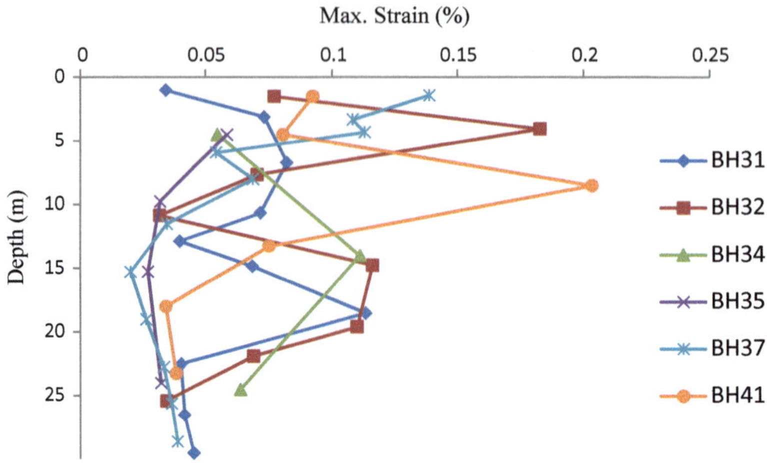

Fig. 12. Maximum strain distribution at Bharalumukh

The spectral acceleration (SA) at surface level for 5% damping are shown from Figs. 13, 14, 15 and 16 for the four sites. The spectral acceleration for all the fours sites are lesser than the spectral acceleration of rocky or hard soil sites (IS: 1893–2002). The value of spectral acceleration has been observed to be low with PGA of 0.152 g of the Sikkim motion because of low energy distribution within the cut-off frequency range.

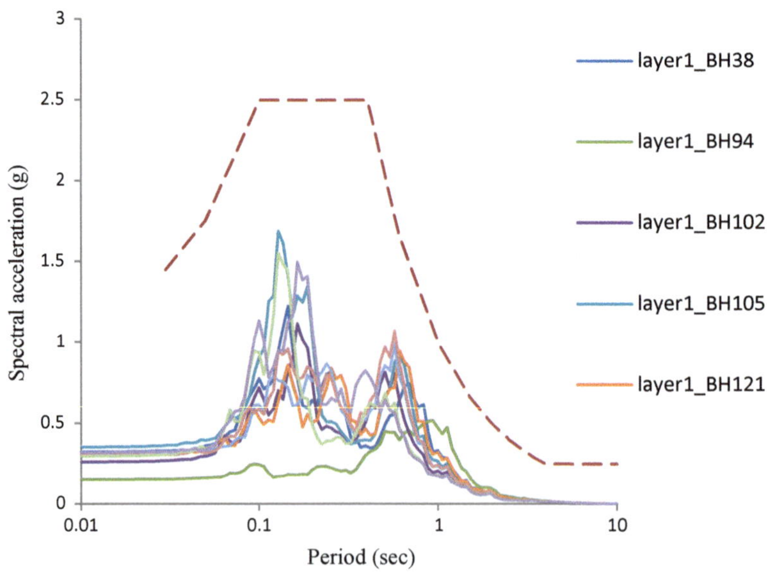

Fig. 13. Spectral acceleration at Beltola

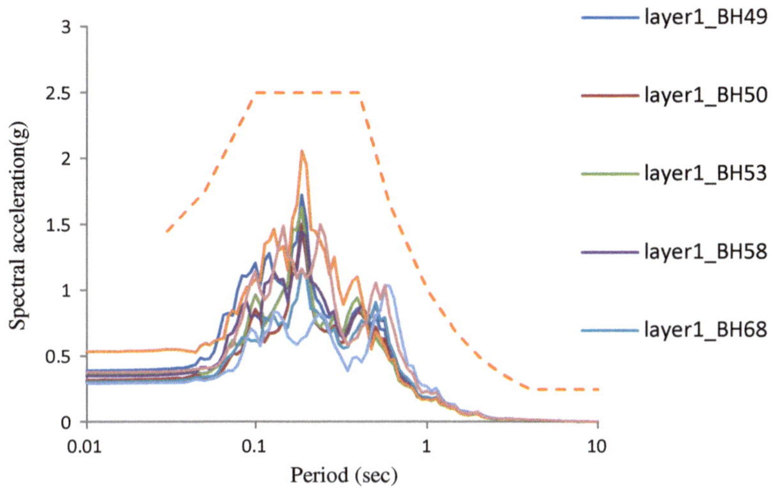

Fig. 14. Spectral acceleration at Panbazar

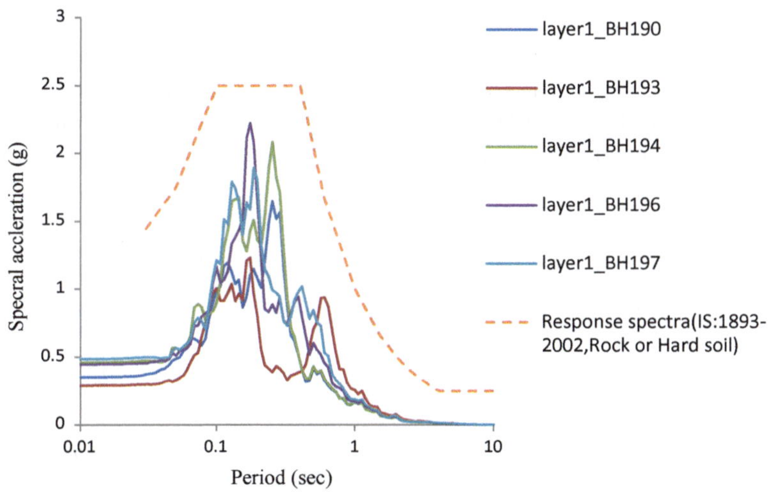

Fig. 15. Spectral acceleration at Lal Ganesh

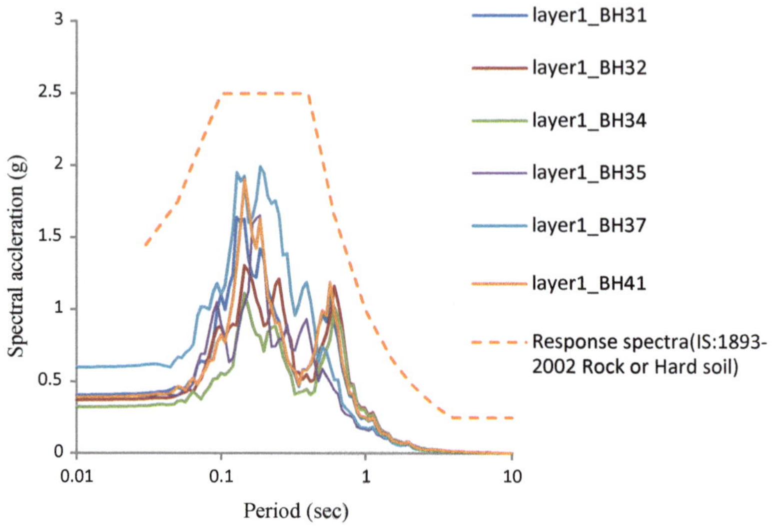

Fig. 16. Spectral acceleration at Bharalumukh

Of the nine borehole sites in Beltola location, two sites of B.H.38 and B.H. 102 are susceptible to liquefaction. The soil layers in between 1.5 m and 2 m from the ground surface are likely to liquefy (FOS<1.0) for B.H.38, whereas liquefaction is likely to occur in the soil column from a depth of 15 m to 16 m from the ground surface for B.H.102 (Fig. 17).

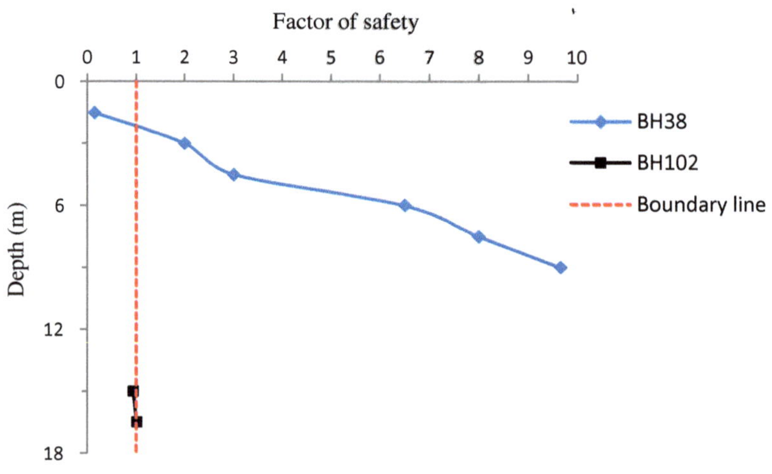

Fig. 17. FOS versus depth at Beltola

Out of the eight boreholes located at Panbazar area, four boreholes viz. B.H.49, B.H.50, B.H.68 and B.H.179 are prone to liquefaction i.e. the F.O.S. <1.0 as shown in Fig. 18. It is seen that liquefaction is likely to occur in the soil column for a depth of 10.5 m to 13.5 m from the ground surface for B.H.49. Similarly depth from 10.5 m to 12 m for B.H.50, from 9 m to 11.5 m for B.H.68 and from 7.5 m to 10.5 m for B.H.179 from the ground surface are prone to liquefaction. In Lal Ganesh area all the five sites are safe against liquefaction (FOS>1.0). Only two locations i.e., B.H.31 and B.H.37 located in Bharalumukh area are likely to liquefy as shown in Fig. 19. Liquefaction would be expected only at 15 m depth from the ground surface for B.H.31and from 9 m to 11 m depth in the soil column are prone to liquefaction for B.H.37.

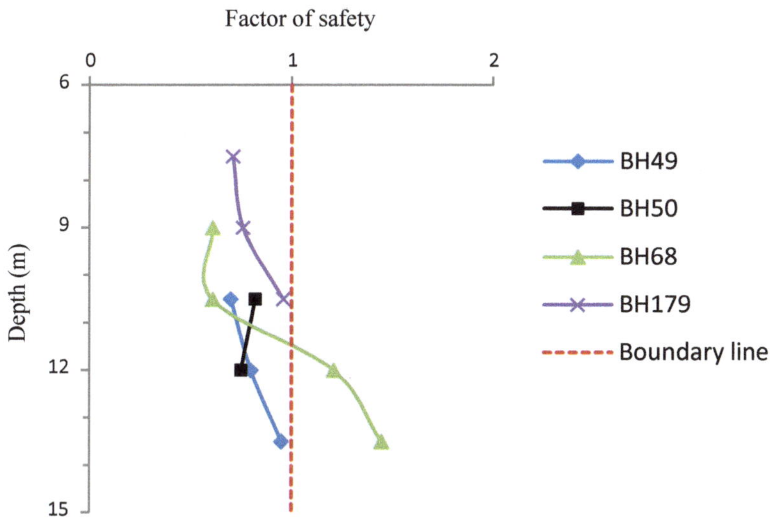

Fig. 18. FOS versus depth at site Panbazar

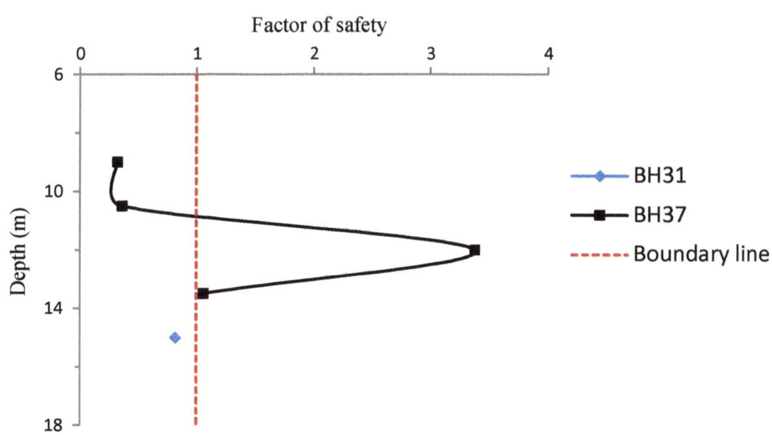

Fig. 19. FOS versus depth at site Bharalumukh

For the other locations in the eastern and southern Guwahati area, similar discussions are not presented in the paper.

Higher maximum strain range is associated with earthquake motion having high energy content. Again for less stiffer soils or soft soils having low SPT-N value or low shear wave velocity, higher maximum strain is released. More dissipation of energy is associated with higher strain levels which lead to the softening of the soil with number of loading cycles. Again less dissipation of energy leads to higher amplification of the PGA at various levels of the soil strata. For the Sikkim motion it is observed that it generates lower cyclic shear strains in the soil sites. This is because its energy content is not very high. Lower cyclic shear strains leads to less dissipation of energy which in turn leads to higher amplification of the PGA at various levels of the soil strata.

GIS based map of factor of safety against liquefaction for all the one hundred twenty (120) boreholes are represented in Fig. 20. After the analysis it has been found that a total of forty two (42) boreholes show a value of factor of safety less than 1.0, showing a sign of possibility of liquefaction. Such important sites are Fatashil ambari, Latashil, Ganeshguri, Chandmari, Panbazar, Hatigaon, Beltola, kahilipara, VIP road, Bamunimaidan, Kalapahar, Sixmile and Bharalumukh.

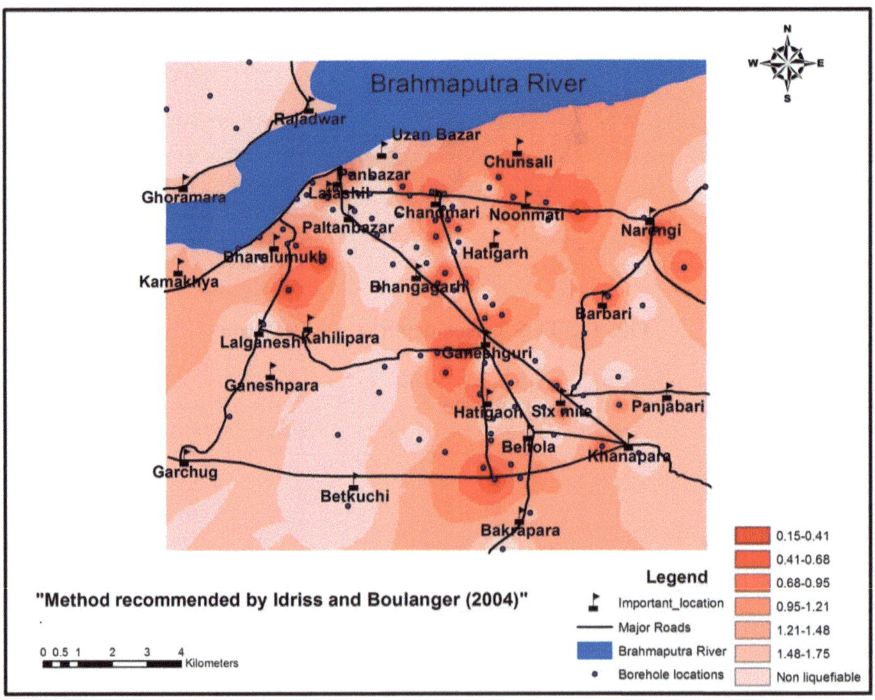

Fig. 20. Liquefaction potential map of southern and eastern part of Guwahati city

There are limitations of one- dimensional equivalent linear ground response analysis. The main limitations are that both the soil and bedrock surface are assumed to extend infinitely in the horizontal direction and all the boundaries are assumed to be horizontal. The response of a soil deposit is caused by SH- waves propagating vertically from the underlying bedrock. The soil layer is considered as a linear visco-elastic material, in which constant shear moduli and damping is used through out each iteration of the analysis.

Further study is necessary specially using non linear method of ground response analysis to get a better picture of the soil response. This work has been mainly carried out to initiate such studies in this seismic prone region and to address the problems caused due to seismic hazards.

5 Conclusions

One-dimensional equivalent linear ground response analysis of one hundred twenty (120) boreholes in Guwahati city using the Sikkim motiom earthquake, showed that the soft soil deposits in areas of Kalapahar, Fatashil Ambari, Jonali, Kachari, Lal Ganesh, Paltanbazar and Ulubari cause strong amplification of seismic motions whereas, Bar Sajai Ghoramara, Hatigaon and Guwahati club areas showed less amplification due to the presence of deeper sandy deposits. Thus, local site geology plays a vital role in the generation of amplified motions at the ground surface. The Fourier amplitude ratio (FAR) at surface level has been obtained in the range of 3.75–9.49 with fundamental frequency range from 1.02 Hz to 3.75 Hz. The spectral acceleration for all the boreholes located at Beltola, Panbazar, Lal Ganesh and Bharalumukh area are lesser than the spectral acceleration of rocky or hard soil sites (IS: 1893–2002). For the Sikkim motion it is observed that it generates lower cyclic shear strains in the soil sites.

Liquefaction potential of the fifty four (54) identified sites in Guwahati city using the surface PGA values showed that a total of forty two (42) boreholes show a value of factor of safety less than 1.0. GIS based map of factor of safety against liquefaction for all the 120 boreholes show that areas Fatashil Ambari, Latashil, Ganeshguri, Chandmari, Panbazar, Hatigaon, Beltola, kahilipara, VIP road, Bamunimaidan, Kalapahar, Sixmile, Bharalumukh, etc. are vulnerable to liquefaction.

Acknowledgments. The Geotechnical data of the two hundred (200) boreholes were taken from a project work given to Assam Engineering College, titled "Liquefaction potential determination of Guwahati city" funded by The Department of Science and Technology, India for Microzonation of Guwahati city. We acknowledge the help and assistance given by DST, India for the study.

References

Boominathan, A., Dodagoudar, G.R., Suganthi, A., Maheswari, U.R.: Seismic hazard assessment of Chennai city considering local site effects. J. Earth Syst. Sci. **117**(S2), 853–863 (2008)

Boore, D.M.: Stochastic simulation of high-frequency ground motions based on seismological models of the radiated spectra. Bull. Seismol. Soc. Am. **73**, 1865–1894 (1983)

Hashash, Y.M.A., Musgrove, M.I., Harmon, J.A., Groholski, D.R., Phillips, C.A., Park, D.: DEEPSOIL version 6.1, User Manual 137 p (2016)

Idriss, I.M., Boulanger, R.W.: Semi-empirical procedures for evaluating liquefaction potential during earthquakes. In: 11th International Conference on Soil Dynamics & Earthquake Engineering (ICSDEE) and the 3rd International Conference on Earthquake Geotechnical Engineering (ICEGE) (2004)

IS (1893–2002). Indian Standard Criteria For Earthquake Resistant Design of Structures, Part 1 General Provisions and Buildings, 5th edn. Bureau of Indian Standards, New Delhi, p. 15

Kramer, S.L.: Geotechnical earthquake engineering, p. 653. Prentice Hall, New Jersey (NJ) (1996)

Kumar, S.S., Krishna, A.M.: Site-specific seismic ground response to different earthquake motions. In: Proceedings of Indian Geotechnical Conference, December 13–15, 2012, Delhi (2012)

Motazedian, D., Atkinson, G.M.: Stochastic finite-fault modeling based on a dynamic corner frequency. Bull. Seismol. Soc. Am. **95**, 995–1010 (2005)

Nath, S.K., Thingbaijam, K.K.S., Raj, A.: Earthquake hazard in Northeast India—a seismic microzonation approach with typical case studies from Sikkim, Himalaya and Guwahati city. J. Earth Syst. Sci. **117**, 809–831 (2008)

Raghukanth, S.T.G., Dash, S.K.: Deterministic seismic scenarios for North East India. J. Seismolog. **14**, 143–167 (2009)

Seed, H.B., Idriss, I.M.: Soil Moduli And Damping Factors for Dynamic Response Analyses. Technical report EERRC-70-10, University of California, Berkeley (1970)

Sharma, B., Rahman, S.K.: Use of GIS based maps for preliminary assessment of subsoil of Guwahati city. J. Geosci. Environ. Protect. **2016**(4), 106–116 (2016)

Vucetic, M., Dobry, R.: The effect of soil plasticity on cyclic response. ASCE Geotech. J. **117**(1), 89–107 (1991)

A Conceptual Model for Estimation on the Strength of Soil Rock Mixture After Grouting Reinforcement

Rentai Liu, Zhuo Zheng(✉), and Qingsong Zhang

Geotechnical and Structural Engineering Research Center, Shandong University, Jinan 250061, Shandong, China
413708318@qq.com

Abstract. Soil rock mixture (SRM), which consists of rock blocks within soft matrix composed of sand, silt and clay, is often regarded as a kind of problematic geology during geotechnical engineering project. Grouting has been used as an effective method to improve mechanical behaviors of SRM. The main goal of this work is a comprehensive study of the physical mechanisms of grouting reinforcement on SRM. For this purpose, laboratory tests are carried out to simulate the process of grouting into SRM. Then, distribution of the grout in SRM is examined to investigate fracturing behavior of the grout suspension. The uniaxial compressive strength of the test samples are measured to investigate the influence of grouting process on the strength of SRM. Different controlling parameters are analyzed, such as volume of block proportion, block count and grouting volume. Based on the analysis, a concept model is proposed which considers three different mechanisms of grouting reinforcement. Finally, an empirical equation is proposed for estimation on the strength of grouted SRM. Results show that the proposed model can reasonably estimate the strength of the test samples.

Keywords: Soil rock mixture · Grouting suspension · Compressive strength Empirical equation

1 Introduction

Soil rock mixture, which consists of rock blocks within soft matrix composed of sand, silt and clay, is a common geological condition during geotechnical engineering project. Medley (1994) proposed the concept of "bimrock", which refers to mixtures of rocks composed of geotechnically significant blocks, within a bonded matrix of finer texture [1]. Soil rock mixture is also incorporated into this classification, and preferred for complex mixtures which include rock blocks surrounded by soil-like matrix material, such as colluvium and glacial tills [2]. In many cases, roil and rock mixture is regarded as a kind of problematic geology, which threatens the security of the construction of engineering project [3, 4].

Grouting has been used as an effective method for problematic geology treatment during construction of underground projects [5–8]. However, the grouting process is complex and the grouting results cannot be easily evaluated. This uncertainty mainly

M. Barman et al. (eds.), *Transportation and Geotechniques: Materials, Sustainability and Climate*, Sustainable Civil Infrastructures,
https://doi.org/10.1007/978-3-319-95768-5_14

attributes to two aspects. Firstly, it is difficult to measure geomechanical properties of the original geological masses [9, 10]. Moreover, the reinforcement mechanism of grouting in soil rock mixture is complex, since the penetration process and interaction between grout and the geological masses is invisible after grout material is injected into the stratum [11–13].

The main goal of this work is a comprehensive study of the physical mechanisms of grouting reinforcement on SRM. For this purpose, laboratory tests are carried out to simulate the process of grouting into SRM. Then, the uniaxial compressive strength (UCS) of the test samples are measured to investigate the influence of grouting process on the strength of SRM. Finally, a concept model and an empirical equation are proposed for estimation on the strength of grouted SRM. Results show that the proposed model can reasonably estimate the strength of the test samples (Fig. 1).

Fig. 1. Test equipment for grouting into SRM

2 Fundamental Mechanics of Soil Rock Mixture

The mechanical property of SRM is influenced by a serious of factors, such as ratio between block and matrix strength, volumetric block proportion (VBP), etc.

One main characteristic of SRM is obvious contrast of strength between the block and the matrix. In this view, the mechanical behavior of SRM is similar to that of bimrock, despite the matrix is weaker for SRM. According to Lindquist [14, 15] and Medley [1], when test samples fail under normal load, the failure surface usually passes tortuously around blocks. Thus, the strength of blocks does not necessarily increase the overall strength of the bimrock. For SRM, the situation is the same. However, when the test samples are under high confining pressure, failure surfaces may directly penetrate blocks, instead of tortuously pass around them.

VBP can significantly influence the mechanical property of SRM [16–21]. When VBP is higher than about 60%, blocks begins to contact each other. A confining pressure may intensify the interlocking behavior between these blocks, which will change the strength and deformation property of SRM dramatically. In this case, the influence of surface roughness of blocks also increases, since a rough surface or irregular shape increases the interlocking behavior.

The mechanical property of SRM is also influenced by the number of blocks (block count). The interface between block and matrix is the weakest part in SRM. When block count increases, the total area of the interface also increase. VBP has no capability of defining the amount of the total area of the interface. While the VBP keeps unchanged, the overall strength of SRM will decease as the block count increases, due to a larger area of interface. On the other hand, a bigger block count may lead to a more uniform distribution of block in SRM, which improves the overall strength of SRM. The total influence of block count should consider both the positive and negative aspect. Thus, block count should also be regarded as a crucial parameter for predicting the strength and deformation properties of SRM. The influences of block count on bimrock are systematically studied by H. Sonmez. However, for SRM, especially with consideration of grouting reinforcement, the influences of block count are not clear and need further investigation.

3 Sample Preparation and Experimental Tests

3.1 Device and Material Preparation

The experiments were performed using a rigid cylindrical steel container. The height of the interior of the container is 25 cm and the diameter is 10 cm. The thickness of the cylinder wall and lids is 1 cm. An injection hole is mounted on the top lid. The bottom lid is punched with five tiny holes with a diameter of 1 mm, which allows water to penetrate, but not soil. A tube is buried into the mixture and connected to the grouting entrance on the top lid before the container is screwed up. The tube is 20 cm long and coincides with the vertical, central axis of the container. There are several holes on the tube, so that the grout suspension can flow into the mixture from different point along the central axis, rather than only from the end of the tube. After the container is filled up with the mixture, it is connected to a pressure pump, and grouting suspension is injected from top of the container.

3.2 Material Preparation

SRM specimen was prepared using soil and gravel at different ratio. Particle size distribution analysis was performed on the mixture, as shown in Fig. 2. The soil is oven-dried at 120 °C for one day before being used for preparing the mixture. The gravel has irregular morphology with a density of 2830 kg/m^3. The mixing and filling process were divided into several times. Each time a constant ratio of soil and gravel were mixed, and cylinder container is filled up layer by layer. The overall compaction degree is controlled by the total amount of mixture filled into the container. The first

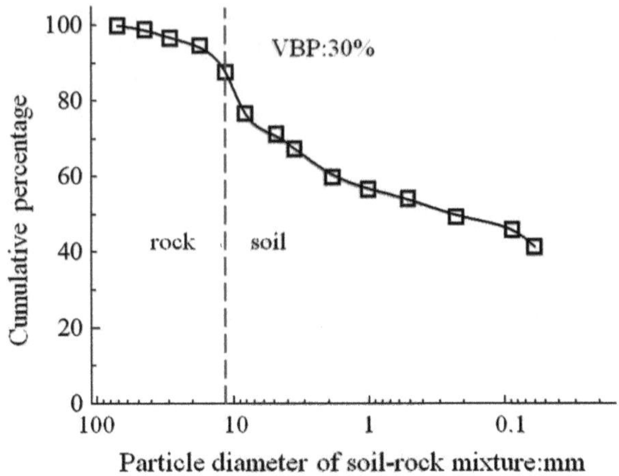

Fig. 2. Grain size accumulation curves for the soil–rock mixtures

layer and last layer are pure soil and compacted tightly. Each of them is 2.5 cm high after compaction. These two layers are expected not to be grouted and will be cut off before compression test. Therefore, the specimen used for compression test is 20 cm high in all, and the diameter is 10 cm.

3.3 Injection

The grout suspension is injected into the cylinder container from the pump. The water-cement ratio is 1:1 for preparation of the grout suspension. During injection, the grout suspension is being stirred persistently in case of sentiment. The pump volume for one time is about 24 ml, and one pump takes 3 s. Thus, the injection rate is considered as 8 ml per second. During injection, water is bleed from the grout suspension and penetrates into the soil, and the grout suspension becomes thicker, so a higher pumping pressure is needed. The termination of the injection is controlled by two standards: the final injection pressure and the total suspension volume injected. If the injection pressure exceeds 2 MPa or the injected suspension volume reaches 1.5 kg, the injection is stopped.

3.4 Compression Test

After injection, the container was kept with the final pressure for 1 h before the specimens were unmolded. The specimens were cured for 28 days at the 100% relative humidity and temperature of 25 ± 5 °C, before used for compression test. The uniaxial compressive test was carried out in accordance with ISRM (1979) at a constant loading rate of 0.05 mm/s. In each test, the loading was continued until full failure of the specimen occurs.

4 Discussion of the Laboratory Test Results

4.1 Grouting Behavior

A comparison of grout distribution in specimen with different VBP was conducted. Results show that, the distribution of blocks significantly influences the grouting behavior, as shown in Fig. 3. For specimen composed of pure soil, there is only one grout vein, and the other area remains ungrouted. The thickness of the grout fracture is influenced by a serious of factors, such as grouting pressure, compaction degree of the soil and the rheological behavior of the grout suspension. For specimen composed of

Fig. 3. Grout distribution in pure soil and SRM

SRM, the grout distribution is much different. The interface between block and soil is a weak place, compared with internal area in the soil. Thus, the grout suspension may penetrate along the boundary of the blocks with priority. The blocks change the original patterns of grout distribution, and cause grout suspension spread into a larger area compared with that in pure soil. Thus, a larger area is reinforced and the overall mechanical property is further improved.

4.2 Influence of Grouting on the Mechanical Property of Soil Rock Mixture

When VBP is high, the distance between blocks is comparatively small. Thus, different blocks become bonded together as a whole after grout have been hardened. In this case, the overall strength of the SRM is improved remarkably. On the other hand, when VBP is low, the distance between blocks is comparatively large. The pressure of the grout suspension should be high enough to overcome the internal stress in the soil, so that it can be fractured by grout suspension. If the pressure is not high enough, the grout suspension cannot always fracture the soil, and the grout suspension is more likely to be confined in a local area around some blocks, and the overall strength of SRM can be comparatively low.

4.3 Failure Pattern and Stress Strain Behavior

Failure patterns of test samples with different VBP are shown in Fig. 4. Before the

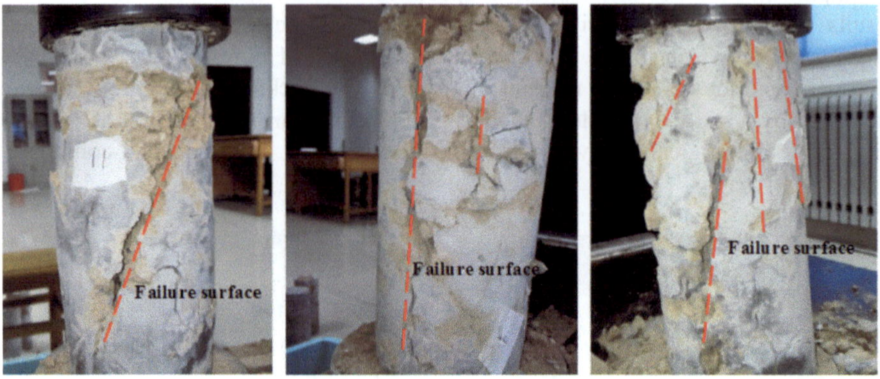

Fig. 4. Failure pattern of test samples with different VBP during uniaxial compressive test

normal stress reaches its peak value, it increase almost linearly with normal stain, and the test sample is considered to be at elastic stage. When VBP is low, the normal stress drops sharply after the peak, and the decreasing rate is very high. The failure surface is continuous and forms almost at once. On the other hand, When VBP is high, the normal stress decreases gradually until it reaches the residual stress. There are several sudden drops which are comparatively small and each of them may correspond to a failure in a local area. The test sample breaks into several parts, but they are not separated completely and can still bear the normal load (Fig. 5).

5 A Conceptual Model for Grouted SRM

Mechanical property of SRM can be reasonably described by Mohr-Coulomb criterion, which has been studied by a serious of researchers [16–21]. Thus, influence of grouting reinforcement can be reflected by improvement of the Mohr-Coulomb parameters, such as cohesive force and friction angle. Three aspects contribute to the improvement of mechanical property of SRM in grouting process: increasing the cohesive force on block boundary, connecting different blocks together, and adding components with higher strength than matrix. With these considerations, a conceptual model for grouting reinforcement on SRM is proposed, as shown in Fig. 6.

A specimen before grouting is represented by Fig. 6a. The blocks with different size distributed randomly in the soil. A potential failure surface is represented by the black dotted line. The specimen after grouting is represented by Fig. 6b. The grout suspension penetrates mainly around the blocks and connects them together. The original failure surface may be prevented, and replaced by a new one with different direction, as shown with the red and black dotted line. Figure 6c represents an ideal

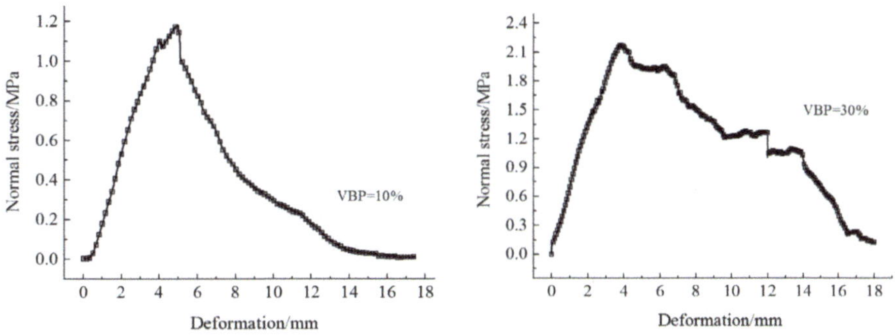

Fig. 5. Stress strain relationship of test samples with different VBP

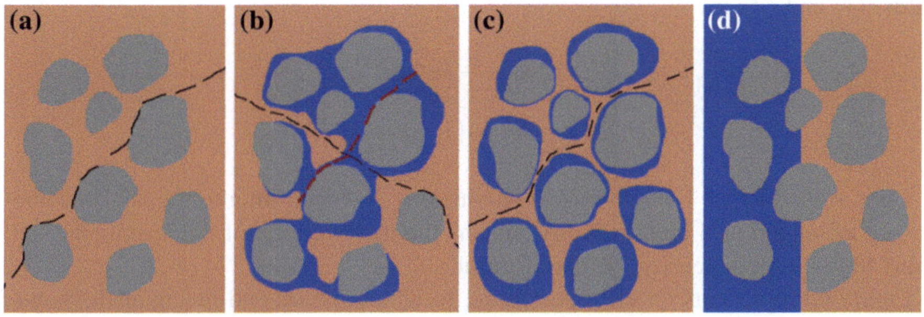

Fig. 6. Grout distribution in SRM: **a** before grouting; **b** a possible distribution after grouting; **c** an extreme case of grout distribution which represents only the first and the third mechanism; **d** an extreme case of grout distribution in which the sample is divided into two distinct parts

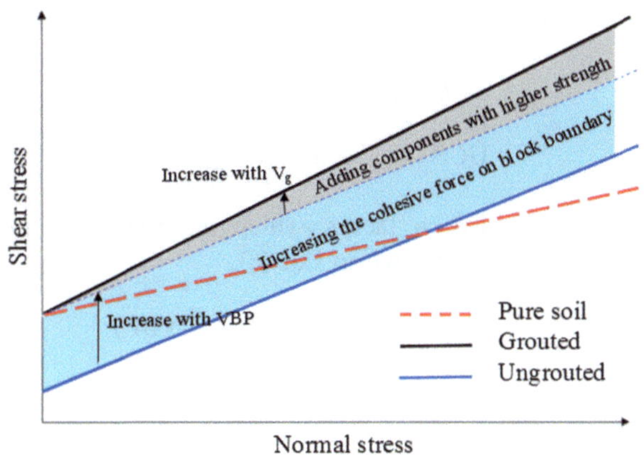

Fig. 7. Comparison of the mechanical behavior before and after grouting reinforcement

condition which only considers the first and the third mechanism. The failure surface may have morphology similar to Fig. 6a, but it propagates through the interior of the soil, rather than along the boundary of the block. Moreover, the boundary of the block is enlarged, and the failure surface becomes more tortuous than that of Fig. 6a, so the overall friction angle also increases. The total improvement is shown in Fig. 7 in terms of Mohr-Coulomb criterion.

The influence of VBP on cohesive has been studied by many researchers. According to their result, A roughly estimation on the cohesion force of SRM as a quarter of that of pure soil is made here. The improvement of internal friction angle is approximately proportional to VBP up to 75%, and the maximum improvement is about 30° but no more than that of the block. Since the injected grout suspension is regarded as an enlargement of blocks, it is reasonable to expect a linear relationship between grout volume and internal friction angle. Based on these rules, a quantitative relationship between the Mohr-Coulomb parameters after grouting is made, as shown in Eqs. (1) and (2).

$$c_{\text{grouted}} = k_1 c_{\text{SRM}} \tag{1}$$

$$\varphi_{\text{grouted}} = \varphi_{\text{SRM}}(1 + k_2 * \text{v}_g/\text{A}) \tag{2}$$

Here, k_1 and k_2 are both empirical parameters, and according to the analysis above, they are estimated as 4 and $2\pi/9$. "A" means the total volume of the specimen, and v_g means the volume taken by the cured grout.

Figure 6d represents an ideal condition, in which the second mechanism shows its maximal influence. The total area is divided into two distinct parts. One of them is sufficiently grouted, which means all the soils are replaced by grout, while the other one is totally ungrouted. A simplification is made here, which assumes that the grouted part is assumed to be composed of pure cement. The strength of the ungrouted part is neglected and the overall strength of the specimen is given in Eqs. (3)–(5).

$$\text{c}_{\text{grouted}} = \text{a} * \text{c}_{\text{cement}} \tag{3}$$

$$\varphi_{\text{grouted}} = \text{a} * \varphi_{\text{cement}} \tag{4}$$

$$\text{UCS}_{\text{grouted}} = \text{a} * \text{UCS}_{\text{cement}} \tag{5}$$

Here, *a* represents a reduction factor, which represents the ratio between the grouted area to the total area, as shown in Eq. (6).

$$\text{a} = \text{V}_g/\text{A} * (1 + \text{VBP})/(1 - \text{VBP}) \tag{6}$$

The real distribution of grout suspension, which is represented by Fig. 6b, is between that of Fig. 6c, d. Thus, Eqs. (1)–(2) and Eqs. (3)–(5) should be regarded as a lower limit and upper limit for the strength of grouted SRM. The actual strength can be calculated based on Eq. (6) by considering another reduction factor *b*, which represents the nonuniformity of grout distribution. This parameter is mainly influenced by block count, *N*.

When block count approaches infinity, b approaches 1, since the grout suspension is considered to be uniformly distributed in the specimen. Thus, the relationship between b and block count is supposed be obey an exponential function, as shown in Eq. (7).

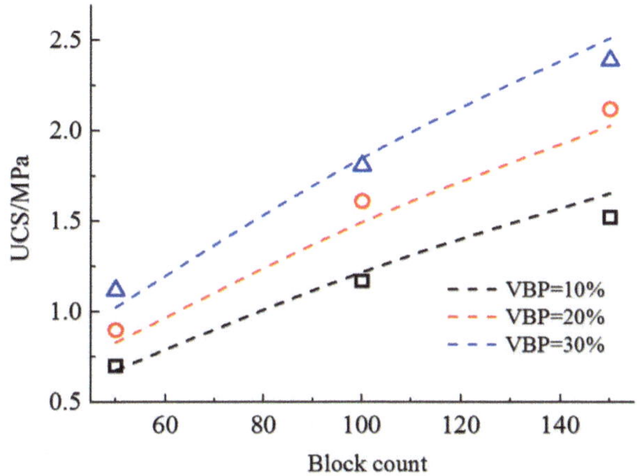

Fig. 8. Comparison of results from empirical equation and laboratory test

$$b = 1 - \exp(\alpha * N) \tag{7}$$

Here, *N* represents block count and α is a fitting parameter. The final equation for estimation of the UCS of grouted specimen is shown below:

$$UCS_{grouted} = V_g/A * (1 + VBP)/(1 - VBP) * [1 - \exp(\alpha * N)] * UCS_{cement} \tag{8}$$

The values of parameters are as follows: V_{inject} is 1000 ml, V_g is 625 ml, A is 1960 ml and UCS_{cement} is 8.8 MPa. α is −0.0044, calculated by regression analysis according to test results. A comparison between calculated value and test value is shown in Fig. 8.

The dashed line represents the proposed empirical equation, and the scatters represent the test value. As can be seen, the UCS increases when block count and VBP increases, and the proposed equation can reasonably estimate the UCS of the test samples.

6 Conclusion

The main goal of this work is a comprehensive study of the physical mechanisms of grouting reinforcement on SRM. Laboratory tests are carried out to simulate the process of grouting into SRM. Then, grouting behavior and mechanical property of the grouted SRM are investigated. Finally, a concept model and empirical equation is proposed for estimation on the strength of grouted SRM. The main conclusions are as follows:

(1) Blocks in soil significantly change the flow patterns of grout suspension. With a high VBP and block count, grout suspension can spread into a larger area compared with that in pure soil. On the other hand, when VBP and block count is low, grout suspension might be confined to some local area around some blocks instead of penetrate into further places, and the overall strength of SRM can be comparatively low.
(2) The strength of the grouted SRM is influenced by VBP. When VBP is low, the normal stress drops sharply after the peak, and the decreasing rate is very high. When VBP is high, the normal stress decreases gradually until it reaches the residual stress. The compressive strength is several times higher than that of low VBP, and the test sample shows ductile property instead of brittle failure.
(3) A conceptual model for grouting reinforcement on SRM is proposed. The model considers three aspects which contribute to the improvement of mechanical property of SRM in grouting process: increasing the cohesive force on block boundary, connecting different blocks together, and adding components with higher strength than matrix. A comparison is made between the test results and calculation from the proposed empirical equation, which shows that the empirical equation can reasonably estimate the UCS of the test samples.

Acknowledgments. This work is support by the national key research and development plan of China (2016YFC0801600).

References

1. Medley, E.: The engineering characterization of melanges and similar block-in-matrix rocks (BIMRock's) (Ph.D. thesis) University of California, Berkeley (1994)
2. Xu, W.J.: Study on meso-structural mechanics (M-SM) characteristics and stability of slope of soil–rock mixtures (S–RM) (Ph.D. diss.) Institute of Geology and Geophysics, Chinese Academy of Science, Beijing (2008) (in Chinese)
3. Xu, W.J., Hu, R.L., Yue, Z.Q., et al.: Genesis and stability of the Zhoujiawan landslide, Three Gorges, China. Bull. Eng. Geol. Environ. **68**, 47–54 (2009)
4. Fragaszy, R.J., Su, J., Siddiqi, F.H., et al.: Modeling strength of sandy gravel. J. Geotech. Eng. **118**(6), 920–935 (1992)
5. Zhang, X., Li, S., Zhang, Q., et al.: Filed test of comprehensive treatment method for high pressure dynamic grouting. J. China Coal. Soc. **35**(08), 1314–1318 (2010) (in Chinese)

6. Zhang, Q., Han, W., Li, S., et al.: Comprehensive grouting treatment for water gushing analysis in limestone breccias fracture zone. Chin. J. Rock Mech. Eng. **31**(12), 2412–2419 (2012) (in Chinese)
7. Ruan, W.-J.: Spreading model of grouting in rock mass fissures based on time-dependent behavior of viscosity of cement-based grouts. Chin. J. Rock Mechan. Eng. **24**(15), 2709–2714 (2005) (in Chinese)
8. Li, S., Liu, R., Zhang, Q., et al.: Research on C-S slurry diffusion mechanism with time-dependent behavior of viscosity. Chin. J. Rock Mech. Eng. **32**(12), 2415–2421 (2013) (in Chinese)
9. Altinsoy, H.: A physical based model investigation for determination of shear strength of block in matrix rocks (M.Sc. thesis) Hacettepe University, Geological Engineering Department, Ankara, Turkey (91 pp.) (2006)
10. Barton, N.: Shear strength criteria for rock, rock joints, rockfill and rock masses: problems and some solutions. J. Rock Mech. Geotech. Eng. **5**, 249–261 (2013)
11. Yang, Z., Hou, K., Guo, T., et al.: Study of column-hemispherical penetration grouting mechanism based on Bingham fluid of time-dependent behavior of viscosity. Rock Soil Mechanics **32**(09), 2697–2703 (2011) (in Chinese)
12. Zhang, X.: Study on mechanism of slurry diffusion and sealing at the process of underground engineering moving water grouting and its application (2011) (in Chinese)
13. Liu, R.: Study on diffusion and plugging mechanism of quick setting cement based slurry in underground dynamic water grouting and its application (Ph. D. thesis) Shan Dong University, Ji Nan (2012) (in Chinese)
14. Lindquist, E.S.: The strength and deformation properties of mélange (Ph.D. thesis) University of California, Berkeley (1994)
15. Lindquist, E.S., Goodman, R.E.: The strength and deformation properties of a physical model mélange. In: Proceedings of 1st North American Rock Mechanics Symposium, Austin, Texas, pp. 843–850 (1994)
16. Vallejo, L.E., Zhou, Y.: The mechanical properties of simulated soil–rock mixtures. In: Proceedings of the 13th International Conference on Soil Mechanics and Foundation Engineering, pp. 365–8. New Delhi, India (1994)
17. Iannacchione, A.T.: Shear strength of saturated clays with floating rock particles. Ph.D. dissertation. University of Pittsburgh (1997)
18. Iannacchione, A.T., Vallejo, L.E.: Shear strength evaluation of clay–rock mixtures. In: Proceedings of GeoDenver, Slope Stability 2000, Denver, Colorado, 3–6 Aug 2000. American Society of Civil Engineers, pp. 209–23 (2000)
19. Yagiz, S.: Brief note on the influence of shape and percentage of gravel on the shear strength of sand and gravel mixture. Bull. Eng. Geol. Environ. **60**(4), 321–323 (2001)
20. Kokusho, T., Hara, T., Hiraoka, R.: Undrained shear strength of granular soils with different particle gradations. J. Geotech. Geoenviron. Eng. ASCE **130**(6), 621–629 (2004)
21. Simoni, A., Houlsby, G.T.: The direct shear strength and dilatancy of sand-gravel mixtures. Geotech. Geol. Eng. **24**, 523–549 (2006)

Immobilization of Pb in Soil Using NZVI Nanoparticles: Effects on Vane Shear Strength

Fuming Liu[1,2], Wan-Huan Zhou[1(✉)], Shuping Yi[2], and Shuaidong Yang[3]

[1] Department of Civil and Environmental Engineering, University of Macau, Macau, China
hannahzhou@umac.mo
[2] Shenzhen Key Laboratory of Soil and Groundwater Pollution Control, South University of Science and Technology of China, Shenzhen, China
[3] Pearl River Hydraulic Research Institute, Pearl River Water Resources Commission, Guangzhou, China

Abstract. In this study, we conducted a comprehensive investigation of the effects of NZVI on the physicochemical and geotechnical properties of Pb(II)-contaminated clayey soil. Soil slurry spiked with lead nitrate were treated with commercial NZVI at a dose of 1%, 5%, and 10% (2–100 mg Fe^0/dry soil g) and further consolidated in a container. A series of Pb sequential extraction procedure, laboratory vane shear test, XRD, and SEM were conducted. Results indicate that the application of NZVI induced a significant decrease in Pb bound to exchangeable and carbonate fractions and a dramatic increase of bound to the Fe–Mn Oxides and the residual fraction. The undrained shear strengths of all soil specimens were determined by vane shear tests. Results show that the introduction of Pb(II) in the soil resulted in a notable decrease in the soil vane shear strength from 16.85 kPa to 7.25 kPa. As expected, the soil vane shear strength was significantly enhanced from 32.41 kPa (1% NZVI) to 69.33 kPa (10% NZVI) after the treatment of NZVI. Simultaneously, the soil particle size of 10% NZVI treated soil specimens were of an evident growth. A new phenomenon that micro level bubble prints in the NZVI treated soil specimen was recorded. These bubbles were presumably generated from the redox of NZVI in the aqueous phase and were particular evidence in the enhancement of soil geotechnical properties. The bulk density and the porosity of Pb-contaminated soil initially decreased then increased with the addition of NZVI. On the whole, the application of NZVI in remediation of Pb-contaminated soil resulted in a dense, compact and reinforced soil.

Keywords: NZVI · Contaminated soil · Physicochemical properties and undrained shear strength

1 Introduction

It is generally known that soil practically offers the essential foundation for production activity. Nonetheless, increasingly soil is being polluted by anthropogenic activities, such as industrial manufacturing, mining, dyeing, finishing and processing industries.

M. Barman et al. (eds.), *Transportation and Geotechniques: Materials, Sustainability and Climate*, Sustainable Civil Infrastructures,
https://doi.org/10.1007/978-3-319-95768-5_15

In order to remedy the contaminated soil, various remediation technologies were developed, for instance, electrokinetic remediation, soil washing, and solvent extraction, stabilization and solidification, permeable reactive barriers, and phytoremediation (Kuppusamy et al. 2016). Thereinto, nanoscale zero valent iron (NZVI) and its iron-mediated nanoparticles are the most commonly employed environmental nanomaterial both in bench and field pilot studies, especially in the United States and Europe (Mueller et al. 2012). Considerable endeavors have been made to improve the efficiency of NZVI including degradation of various pollutants, mobility in the subsurface, revealing the fate and impacts of NZVI on ecology. And they are still under continuous development since its introduction in the later of the past century.

For the NZVI nanoparticles involved soil remediation, on the one hand, from the view of environmental remediation, the purpose of introducing NZVI into subsurface is to remove or stabilize/immobilize the pollutants so that to decrease the potential threaten on ecology. On the other hand, from the view of civil engineering, both pollutants and injected NZVI are nonnative for original soil. With respect to the follow-up infrastructure and constructions on the brownfields and reuse of contaminated soil (such as backfill), the geotechnical properties of remedied contaminated soil are critical data for civil engineering. Consequently, whether the geotechnical properties of soil will change after the implementation of NZVI employed remediation technologies, and how will the NZVI effect on the geotechnical properties of contaminated soil? To the best of our knowledge, limited literatures revealed the effects of NZVI on the geotechnical properties of the contaminated soil.

In this study, we presented an investigation of the stabilization of heavy metal by NZVI in soil and its effects on the undrained shear strength. Particularly, lead (Pb), a heavy metal of well-known toxicity, prevalence and responsible for liver and kidney damage, anemia, infertility, and mental retardation among other consequences, is employed as an artificial contaminant (Kim and Williams 2017). The behavior of NZVI was focused to evaluate the effectiveness in improving the geotechnical properties of soil.

2 Methodology

2.1 Soil Preparation

The soil samples were collected from a construction site in Areia Preta, Macau (N22° 12′35.84″, E113°33′27.96″). Guided by the Unified Soil Classification System (USCS, ASTM D4318), the Atterberg limits of the preconcerted soil were found to be 61, 32, and 29 for the liquid limit, the plastic limit, and the plasticity index, respectively (Stevens 1982). The particle size distribution obtained from the Malvern Particle Size Analyzer shown that the D60, D30, and D10 values of the pristine soil were 8.5, 3.08 and 1.01 mm, respectively (USCS, ASTM D6913). The soil was evaluated as MH (USCS, ASTM D2487), silt with high plasticity.

To obtain a relatively homogeneous soil slurry, the soil was initially severed into small bulks and immersed in water for two days. Then the soil slurry was sieved at 0.425 mm and stirred for homogenization. For the preparation of Pb-contaminated soil specimen, a certain of lead nitrate ($Pb(NO_3)_2$) was added to the soil slurry under continuous agitation. The spiked soil was aged for two weeks for the equilibrium of adsorption and desorption, then further consolidated in a container (Indraratna and Redana 1998). Before poured the soil slurry into the container, a certain of water was added to get the water content of the slurry at 2 times of its liquid limit, then stirred it thoroughly. The overburden pressure of the soil slurry in the container was controlled at approximately 200 kPa. The soil slurry was consolidated until the sedimentation equilibrium was reached, then tailed the obtained soil clod into soil specimens (height: 84 mm, diameter: 38 mm). As for the NZVI treated soil, 1%, 5%, and 10% NZVI (mass ratio of the lead nitrate spiked soil) were added and stirred for homogenization before undergoing the consolidation, then follow the same procedure to get a relatively stout soil sample. NZVI was purchased from Shanghai Xiangtian nano materials co., LTD and used as received. The average diameter of NZVI is about 50 nm, with the specific area of 30 m^2/g.

2.2 Testing Method

Laboratory vane shear test (Wykeham Farrance Laboratory Vane Apparatus, engineering LTD.) was employed to evaluate the undrained shear strength of natural soil specimen, polluted soil specimen and NZVI treated soil specimens. All tests were implemented in a circular tube with the soil specimens loaded inside. The vane blades used in the test are adjusted according to the hardness degree of the specimen. Two kinds of vane blade with the same width of 12.7 mm yet at the different height of 19 mm and 12.7 mm were employed. For all laboratory vane shear tests, the insertion depth of vane blade was maintained at approximately 1.5 cm, from the surface of the soil specimen to the cusp of the blade. The rotation rate of the vane during testing was kept at approximately 360°/h. The undrained shear strength was calculated by the method reported by (Park et al. 2015).

2.3 Sequential Extraction Procedure

The availability of Pb in soil specimens was analyzed by applying the modified Sequential Extraction Procedure that was proposed by Tessier et al. (1979). Extraction solutions with increasing acidic strength and thermometric strength were sequential applied. As a rule, sequential extraction procedure was carried out in the order of exchangeable (EX), carbonate-bound (CB), Fe/Mn oxides-bound (OX), organic matter bound (OM), and residual (RS) fraction. The Pb concentration of supernatant liquor in each sequential extraction procedure was measured by Inductively Coupled Plasma Mass Spectrometry (ICP-MS, Agilent 7700X).

2.4 Characterization

The mineralogy of the soil is characterized by X-ray diffraction (XRD) on Rigaku Smartlab (anode Cu, in configuration $\theta/2\theta$, with an acceleration tension of 45 kV and a current of 200 mA), at a scanning rate of 2.0°/min. The microstructure of soil specimens was depicted by scanning electron microscopy (SEM, Zeiss Merlin) after undergoing vacuum dry oven at 35 °C for 12 h. The soil particle size distribution was characterized by a Laser Particle Size Analyser (Malvern, Mastersizer 3000) with distilled deionized water as the dispersant and ultrasound assisted. Mercury Intrusion Porosimetry (AutoPore IV 9510 Mercury Porosimeter, Micromeritics Instruments Co., USA) was used to describe the pore structure characteristics of soil. The soil was proceeded vacuum freeze-drying for 24 h prior to the MIP measurements.

3 Results and Discussion

3.1 Metal Immobilization

It has been widely accepted that the mechanism of NZVI for immobilization of Pb in the aqueous system was controlled by sorption process, reduction of Pb^{2+} to Pb^0, and (co)precipitation (Gil-Diaz et al. 2017). In this study, the effects of NZVI application on the immobilization of Pb in soil specimens were examined by comparing the distribution of lead concentration in the different soil fractions.

Figure 1 shows concentration distribution tendency of lead retained in the soil specimens, obtained by Sequential Extraction Procedure. The lead concentration in the exchangeable (EX), carbonate-bound (CB), organic matter-bound (OM) as well as the residual fractions significantly increased with the introduction of lead nitrate. As expected, the lead in more easily available friction, namely in exchangeable and

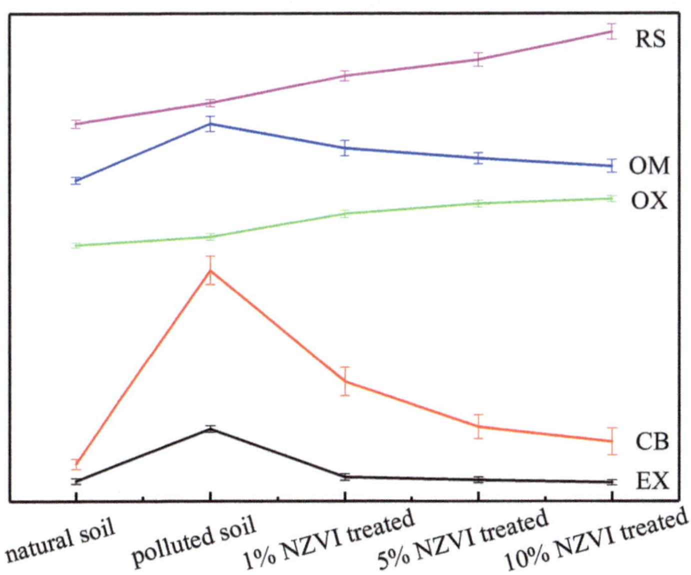

Fig. 1. Sequential relative concentration distribution tendency of Pb in the soil specimens (EX, exchangeable fraction, ×10 ng/g; CB, carbonate-bound fraction, $\times 10^3$ ng/g; OX, Fe/Mn oxides-bound fraction, $\times 10^6$ ng/g; OM, organic matter-bound fraction, $\times 10^5$ ng/g; RS, residual fraction, $\times 10^4$ ng/g). Error bars represent relative standard deviations (n = 2)

carbonate-bound fractions, evidently decreased by the introduction of NZVI. With the increase of NZVI amount, a diminishing concentration of Pb in EX, CB, and OM fractions was observed. For these more easily available frictions, they are of high leachability and bioaccessibility in all fractions (Zong et al. 2016). This is, the application of NZVI can efficaciously immobilize the lead in soil and lower the risk to humans and environmental organism. By contrast, the lead in Fe/Mn oxides-bound and residual fractions increased with the addition of NZVI. It is reasonable because of the interfusion of ions, and part of Fe/Mn oxides-bounded lead converted to a more strongly bound complex. Moreover, part of decomposition of organics by the NZVI resulted in a descending concentration of lead in the organic-bound fraction. In all cases, most leads were bounded by organic matter and Fe/Mn oxides then followed with residual fraction, exchangeable fraction, and carbonate-bound fraction. On the whole, the relative proportion of lead in soil specimens hardly changed after the introduction of NZVI.

3.2 Undrained Shear Strength

In order to evaluate the effects of NZVI on the soil geotechnical properties, the classical soil geotechnical parameter, namely undrained shear strength (S_u), was involved in this study. Primarily, shear resistance of soil flows from the friction and interlocking of soil particles, and possibly cementation or bonding at particle contacts. The result of vane shear test is shown in Fig. 2. It is revealed that the S_u of natural soil decreased sharply with the introduction of lead nitrate, from 16.82 kPa to 7.25 kPa. Generally, it's a synthetic result of the changes of physicochemical properties of soil particles, including

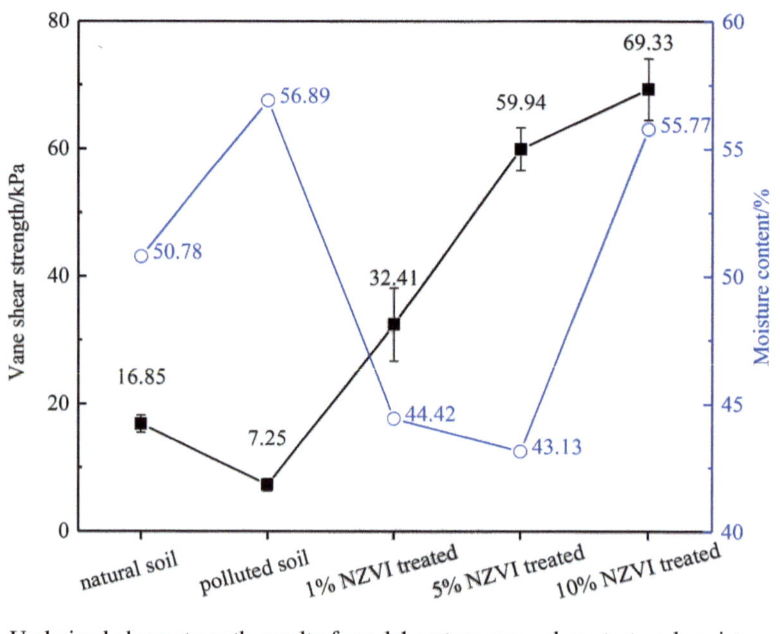

Fig. 2. Undrained shear strength results from laboratory vane shear test and moisture content

water content, framework structure, dielectric constant, the structure of diffuse double layer and pore fluid viscosity (Ratnaweera and Meegoda 2006). Interestingly, we found that there were plenty of milli bubbles equably appeared in the lead-contaminated soil specimen (after unloading from the container). However, little literature reported this particular phenomenon. Speculation is raised for these milli bubbles that they were generated during the mixing procedure and were detained in the lead soil specimen. Moreover, decomposition of soil particle bulks and organics resulted by the introduction of lead nitrate may also produce a certain of bubbles. Thus, the high moisture content, incompact soil particles coupled with the porous structure of lead-contaminated soil induced a decrease of undrained shear strength.

On a positive note, the S_u of lead-contaminated soil markedly got improved via the treatment of NZVI. The S_u of soil increased with the increase of the amount of added NZVI but tardily increased when the amount of added NZVI is 10%. The S_u of soil enhanced up to 69.33 kPa when the amount of added NZVI is at 10% soil dry mass. At the same time, the moisture content of the soil specimens decreased with the increase of the added amount of the NZVI, but the moisture content of the 10% NZVI treated soil singularly mounted up to 55.77%. To some extent, the relationship between soil moisture content and shear strength seems irrelevant as the introduction of NZVI. For the NZVI treated soil, plenty of tiny fissures and microbubbles were present in the appearance of the soil slurry after the introduction of NZVI. Accordingly, a large quantity of microbubbles remained in the soil specimen after unloading from the container. This abnormal phenomenon provides an unequivocal evidence of interaction between NZVI and the clayey slurry.

3.3 X-ray Diffraction (XRD)

The X-ray diffraction (XRD) patterns of the soil specimens are shown in Fig. 3. Analysis of XRD data indicates that the selected soil is primarily composed of quartz (PDF#85-1054, ICSD Minerals), laumontite $Ca_4(Al_8Si_{16}O_{48}(H_2O)_{18}$ (PDF#80-0150, ICSD Minerals), and greenalite $Fe_6Si_4O_{10}(H_2O)_{18}$ (PDF#72-0177, ICSD Minerals). It

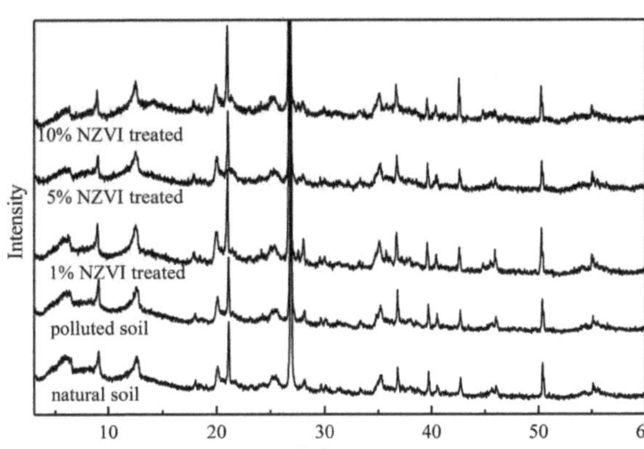

Fig. 3. X-ray diffraction (XRD) patterns of the soil specimens

could be seen that the main characteristic peaks revealed slight discordance among the soil specimens, implying that distortion occurred to the lattice of minerals after the introduction of NZVI. It is assumed that the frameworks of a certain of mineral crystals were doped by Fe(II), Fe(III) and Pb(II) that of larger atomic radius, resulting in the slight but appreciable changes of lattice parameters (Castaldi et al. 2008). The tiny difference in weak characteristic peaks could be perceived at 2-Theta of 27.6°, 35.7°, 36.6°, and 42.6°, which can be credited to the neonatal iron mineral crystals of ferrosilite $FeSiO_3$, hematite Fe_2O_3, goethite FeOOH, and magnetite Fe_3O_4, respectively (Huang et al. 2013). As a result of the complexity of the soil mineral composition, the characteristic peaks of other different iron minerals such as maghemite (γ-Fe_2O_3) are concealed. The presence of deuterogenic different iron minerals will constitute the highly-disordered multilayer of amorphous core shell construction on the surface of NZVI and soil particles with a thickness of approximately several nanometers (Arancibia-Miranda et al. 2014).

For the NZVI in the redox system, one the one hand, the reduction of Pb^{2+} will be suppressed due to isolation resulted from the multilayer core shell construction; On the other hand, the corrosion byproducts can greatly promote the co-precipitation and adsorption of Pb^{2+}, and improve the removal capability of NZVI. For the soil particles, they were efficaciously bonded together through the deuterogenic different iron minerals to form a firmer skeleton structure and improve the soil geotechnical properties. However, the characteristic peaks associated with reduced lead and related lead oxides were not clearly presented in the diffractogram of NZVI treated soil samples may in consequence of (1) the minerals in the clayey soil are too complex which dwarf the characteristic peaks associated with reduced lead and related lead oxides; (2) the Pb is of trace amount in soil as well as (3) the crystalline structures of reduced lead and related lead oxides are amorphous.

3.4 Microstructure and Particle Size

Changes of soil physicochemical properties and geotechnical properties are always instantiated in the microstructure of soil. In this study, the microstructure of soil was recorded to clarify the effect of NZVI application on the soil geotechnical properties. As was mentioned before, milli bubbles in the body of Pb-polluted soil specimen could be attributed to the decomposition of soil particles by the introduction of lead nitrate (Fig. 4a). A novel phenomenon was found in the NZVI treated soil specimens. It is found that NZVI aggregations were evident even after acutely agitation of the soil slurry and NZVI. Moreover, a mass of equal microbubbles was generated both in the soil slurry and soil specimens (namely before and after the consolidation). NZVI is easily oxidized to ferrous iron in aerobic condition and protons are consumed in the process:

$$Fe^0 + O_2 + 2H^+ \rightarrow Fe^{2+} + H_2O_2 \tag{1}$$

In the reaction (1), hydroxyl radicals ($\cdot$OH) spontaneously generated as the combination of the produced H_2O_2 and Fe^{2+} (known as Fenton reaction or Fenton-like

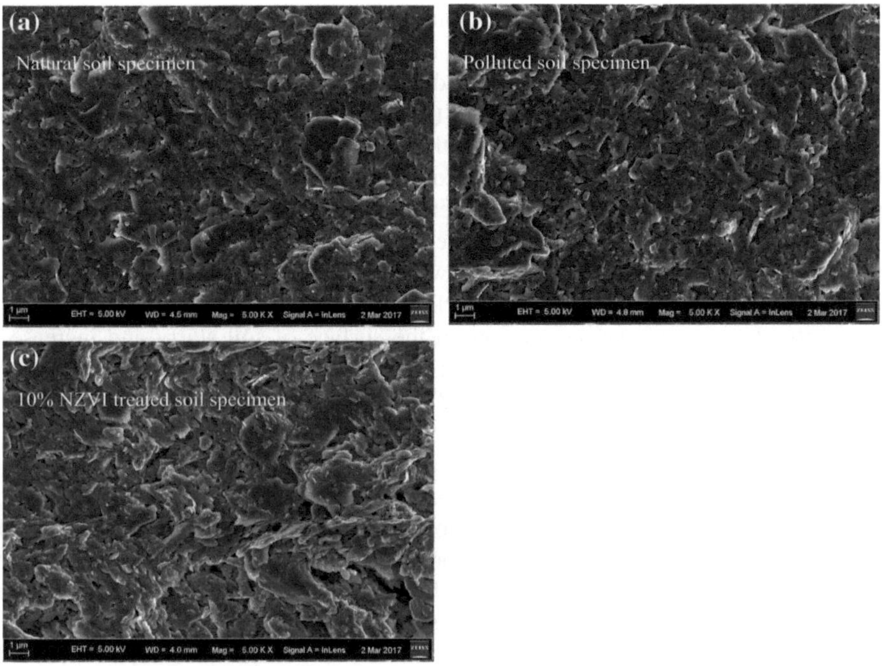

Fig. 4. Microstructure of soil specimens (after unloading from container): **a** natural soil specimen; **b** polluted soil specimen; **c** 10% NZVI treated soil

reaction). When oxygen is exhausted or in the absence of oxygen, NZVI tends to experience a rapid oxidation phase with the generation of hydrogen gas:

$$Fe^0 + 2H_2O \rightarrow Fe^{2+} + 2OH^- + H_2 \quad (2)$$

The hydroxyl radicals ($\cdot OH$) possess strong and non-selective oxidizing capability of mineralization of most organic compounds, the ferrous will be oxidized to ferric:

$$Fe^{2+} + H_2O_2 \rightarrow Fe^{3+} + \cdot OH + OH^- \quad (3)$$

In addition to that, the oxidative ferric iron ions can be reduced by zero-valent iron to provide more ferrous, and more associated hydroxyl radicals to accelerate the degradation of organic compounds:

$$2Fe^{3+} + Fe^0 \rightarrow 3Fe^{3+} \quad (4)$$

The established iron redox cycling promotes the generation of ferric and ferrous ions. Meanwhile, iron oxide/hydroxide will form on the surface of NZVI as the erosion of NZVI. Iron oxides or hydroxides are capable of adsorption of contaminants on the surface of NZVI (Noubactep 2008). Under particular environment, these hydroxides would through dehydration and crystallization to be transformed into various less adsorptive oxides such as FeO, FeOOH. On the other hand, as the deuterogenic ferric and ferrous temporarily produced in the iron redox system are only stable under specific condition, they are easily oxidized to insoluble iron species such as Fe_3O_4 (maghemite), Fe_3O_4 (magnetite), FeOOH (lepidocrocite), or goethite (Lefevre et al. 2016). Evidently, hydrogen generated by the oxidation of NZVI and the subsequent oxidative transformation was obligated to these microbubbles. In short-term, the microbubbles generated from the redox reaction of NZVI may lead to a volume change combined with a go-out of gas which depends on the air permeability of the soil. In long term, a state of equilibrium will occur with microbubbles in the soil body that may leave a series of residual channeled network of voids for the drainage boundary. As a result of the decrease of surface tension on the surface of soil slurry and increase of drag force between soil particles, massive fissures were grown on the surface of NZVI treated soil slurry. With the increase of NZVI amount, the microbubbles and fissures are clearer.

The soil specimens (after the consolidation) were also characterized by SEM to classify the physical changes of soil particles, as is shown in Fig. 4. In all cases, the soil particles stuck together in anomalous clumps after undergoing the consolidation in the container, with particle size distributes from 400 nm to 1 mm. All the added NZVI have been oxidized to neonatal iron mineral during the consolidation in the container, no NZVI particle was shown on the soil particles. The decomposition of soil particles by lead nitrate leads to the increase of clay and silt clay fraction, scilicet, reduced the soil particle size. On the other hand, it is found that with the increase of NZVI, the soil particle size gravitates towards to increase, as shown in Fig. 5. However, the effect of NZVI on the soil particle size is not that evident until the added amount of NZVI is 10% of the dry soil.

Figure 6 shown the bulk density and porosity of soil specimens determined by MIP in the range from 3 nm to 360 μm. Results indicate that both the bulk density and the porosity of soil significantly decreased as the introduction of Lead Nitrate and NZVI.

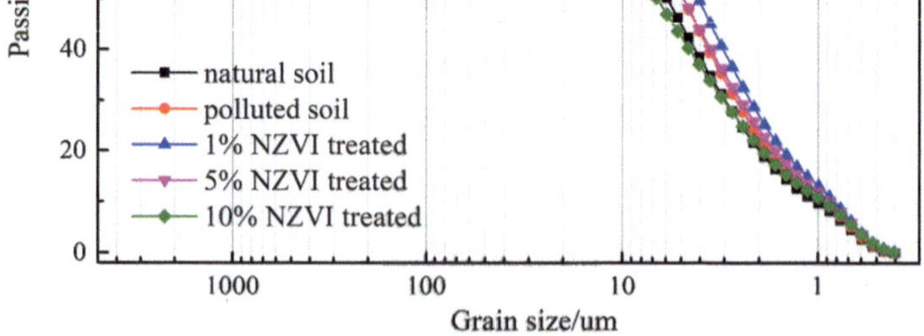

Fig. 5. Particle size distribution of the clayey soil under different treatment conditions

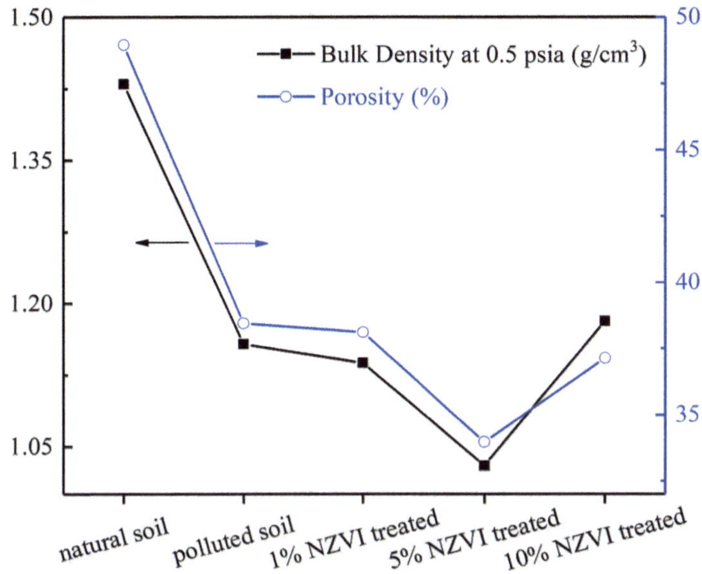

Fig. 6. Bulk density and porosity of soil specimens determined by MIP in the range from 3 nm to 360 μm

Particularly, they present an identical tendency for the treatment by NZVI: the bulk density and the porosity of Pb-contaminated soil initially decreased then increased with the addition of NZVI. In other word, the application of NZVI in remediation of Pb-contaminated soil resulted in a dense, compact and reinforced soil. The decreased porosity of NZVI treated soil can ascribed to the rapid aggregation and oxidation of bare NZVI combined with clogging phenomena in the soil pore (Jiemvarangkul et al. 2011).

Based on the analysis of soil particle size, pore characteristics and theory of redox reaction of NZVI in the soil slurry, a conclusion could be drawn that the increase of undrained shear strength is primarily assigned to the NZVI induced neonatal iron minerals precipitation on the soil particles and bonding them together. At the same time, the generated bubbles in the soil body may provide some preferential channels for draining water out and partly improved the soil undrained shear strength (Acharya et al. 2016). The introduction of NZVI in soil resulted in a dense, compact and reinforced soil.

4 Conclusions

In this research, a laboratory study was performed to investigate the effects of NZVI on the immobilization of lead and the soil undrained shear strength. Results indicate that the application of NZVI can efficaciously lower the lead in the exchangeable (EX), carbonate-bound (CB), and organic matter-bound (OM) fractions in soil. The introduction of Pb(II) in the soil resulted in a visible decrease of vane shear strength from 16.85 kPa to 7.25 kPa. Simultaneously, the moisture content of soil specimen slightly increased by the introduction of lead.

On a positive note, several positive effects were observed by the introduction of NZVI in the contaminated soil. With the increase of added NZVI: (1) the shear strength of NZVI treated soil was markedly improved from 32.41 kPa (1% NZVI treated) to 69.33 kPa (10% NZVI treated); (2) a significant decrease of Pb bound to exchangeable and carbonate fractions and a dramatic increase of bound to Fe–Mn Oxides and residual fraction were verified; (3) the soil particle size of the 10% NZVI treated soil specimen was of an evident growth; (4) the equal micro-fissures and bubbles appeared on the soil slurry and in the soil specimens.

The results of XRD and soil particle size analysis revealed that increase of undrained shear strength is primarily credited to the precipitation of NZVI induced neonatal iron minerals on the soil particles and the consequent bonding effects among them. Besides, the generated bubbles in the soil body may provide some preferential channels for water drainage and therefore improve the soil shear strength. All in all, the application of NZVI in remediation of Pb-contaminated soil resulted in a dense, compact and reinforced soil.

It shall be noted that further research is ongoing to investigate the mechanism of the strength enhancement during the NZVI treatment. This is important for environmental and geotechnical engineering in the projects that require both the remediation of metals-contaminated soil and reuse of the contaminated site for constructions.

Acknowledgements. The authors wish to thank the financial support from the Macau Science and Technology Development Fund (FDCT) (125/2014/A3), the University of Macau Research Fund (MYRG2017-00198-FST), the National Natural Science Foundation of China (Grant No. 51508585), Shenzhen Peacock Plan (KQTD2016022619584022), and Key Laboratory of Soil and Groundwater Pollution Control of Shenzhen City, South University of Science and Technology of China.

References

Acharya, M.P., Hendry, M.T., Martin, C.D.: Effect of gas bubbles on pore pressure response in peat beneath a railway embankment. Can. Geotech. J. (2016). https://doi.org/10.1139/cgj-2015-0122

Arancibia-Miranda, N., Baltazar, S.E., Garcia, A., Romero, A.H., Rubio, M.A., Altbir, D.: Lead removal by nano-scale zero valent iron: surface analysis and pH effect. Mater. Res. Bull. J. (2014). https://doi.org/10.1016/j.materresbull.2014.07.045

Castaldi, P., Santona, L., Enzo, S., Melis, P.: Sorption processes and XRD analysis of a natural zeolite exchanged with Pb2+, Cd2+and Zn2+ cations. Hazard. Mater. J. (2008). https://doi.org/10.1016/j.jhazmat.2007.12.040

Gil-Diaz, M., Pinilla, P., Alonso, J., Lobo, M.C.: Viability of a nanoremediation process in single or multi-metal(loid) contaminated soils. Hazard. Mater. J (2017). https://doi.org/10.1016/j.jhazmat.2016.09.071

Huang, P.P., Ye, Z.F., Xie, W.M., Chen, Q., Li, J., Xu, Z.C., Yao, M.S.: Rapid magnetic removal of aqueous heavy metals and their relevant mechanisms using nanoscale zero valent iron (nZVI) particles. Water Res. J. (2013). https://doi.org/10.1016/j.watres.2013.01.054

Indraratna, B., Redana, I.W.: Laboratory determination of smear zone due to vertical drain installation. J. Geotech. Geoenviron. Eng. (1998). https://doi.org/10.1061/(asce)1090-0241(1998)124:2(180)

Jiemvarangkul, P., Zhang, W.X., Lien, H.L.: Enhanced transport of polyelectrolyte stabilized nanoscale zero-valent iron (nzvi) in porous media. Chem. Eng. J. (2011). https://doi.org/10.1016/j.cej.2011.02.065

Stevens, J.: Unified soil classification system. Civ. Eng. ASCE **52**, 61–62 (1982)

Kim, M.A., Williams, K.A.: Lead Levels in landfill areas and childhood exposure: an integrative review. Public Health Nurs. J (2017). https://doi.org/10.1111/phn.12249

Kuppusamy, S., Palanisami, T., Megharaj, M., Venkateswarlu, K., Naidu, R.: In-situ remediation approaches for the management of contaminated sites: a comprehensive overview. Rev. Environ. Contam. Toxicol. **236**(18), 1–115 (2016). https://doi.org/10.1007/978-3-319-20013-2_1

Lefevre, E., Bossa, N., Wiesner, M.R., Gunsch, C.K.: A review of the environmental implications of in situ remediation by nanoscale zero valent iron (nZVI): behavior, transport and impacts on microbial communities. Sci. Total Environ. **565**, 889 (2016). https://doi.org/10.1016/j.scitotenv.2016.02.003

Mueller, N.C., Braun, J., Bruns, J., Cernik, M., Rissing, P., Rickerby, D., Nowack, B.: Application of nanoscale zero valent iron (NZVI) for groundwater remediation in Europe. Environ. Sci. Pollut. Res. J. (2012). https://doi.org//10.1007/s11356-011-0576-3

Noubactep, C.: A critical review on the process of contaminant removal in Fe^0–H^2O systems. Environ. Technol. **29**(8), 909–920 (2008). https://doi.org/10.1080/09593330802131602

Park, D., Kutter, B.L., DeJong, J.T.: Effects of thixotropy and cement content on the sensitivity of soft remolded clay. J. Geotech. Geoenviron. Eng. **141**(2), 04014095 (2015). https://doi.org/10.1061/(ASCE)GT.1943-5606.0001221

Ratnaweera, P., Meegoda, J.N.: Shear strength and stress-strain behavior of contaminated soils. Geotech. Test. J. (2006) https://doi.org/10.1520/gtj20150184

Tessier, A., Campbell, P.G.C., Bisson, M.: Sequential extraction procedure for the speciation of particulate trace-metals. Anal. Chem. J. (1979). https://doi.org/10.1021/ac50043a017

Zong, Y.T., Xiao, Q., Lu, S.G.: Chemical fraction, leachability, and bioaccessibility of heavy metals in contaminated soils, Northeast China. J. Environ. Sci. Pollut. Res. (2016). https://doi.org/10.1007/s11356-016-7598-9

Author Index

M. Barman et al. (eds.), *Transportation and Geotechniques: Materials, Sustainability and Climate,* Sustainable Civil Infrastructures,
https://doi.org/10.1007/978-3-319-95768-5

MIX
Papier aus verantwortungsvollen Quellen
Paper from responsible sources
FSC® C105338

If you have any concerns about our products,
you can contact us on
ProductSafety@springernature.com

In case Publisher is established outside the EU,
the EU authorized representative is:
Springer Nature Customer Service Center GmbH
Europaplatz 3, 69115 Heidelberg, Germany

Printed by Libri Plureos GmbH
in Hamburg, Germany